P9-AQA-713

CORRELATION AND CAUSALITY

CORRELATION AND CAUSALITY

DAVID A. KENNY
University of Connecticut

A WILEY-INTERSCIENCE PUBLICATION

JOHN WILEY & SONS, New York
Chichester · Brisbane · Toronto

Library of Congress Cataloging in Publication Data

Kenny, David A 1946-
Correlation and causality.

"A Wiley-Interscience publication."
Bibliography: p.
Includes index.
1. Social sciences-Statistical methods. 2. Social sciences-Mathematical models. 3. Correlation (Statistics) I. Title. [DNLM: 1. Psychometrics. 2. Sociometric technics. 3. Models, Theoretical. HM48 K36c]
HA29.K429 330'.1'82 79-4855
ISBN 0-471-02439-2

Printed in the United States of America

10 9 8 7 6 5 4 3 2

To
Mary Ellen
and My Parents

Preface

Researchers in the social sciences often require reference books to aid them in the computation and interpretation of statistics. These books are usually organized around a set of statistical tools and give extensive detail to the formulas used in estimation. Researchers also use computer packages to compute these statistics. A simple, one-line command will give birth to a multiple regression, factor analysis, or analysis of variance.

On the bookshelves next to the statistical texts and computer manuals are the books that summarize and interpret the field. These texts of substantive theory may occasionally present statistical summaries of data, but data play a small role in such texts. To some extent, theory and data rarely touch each other in the social sciences. To bridge the gap, over the past ten or so years, an area called causal or structural analysis has developed. This area takes as its task the explication of a statistical method in terms of how the method relates to substantive theory.

This text is a general introduction to the topic of structural analysis. It is an introduction because it presumes no previous acquaintance with causal analysis. It is general because it covers all the standard, as well as a few nonstandard, statistical procedures. Since the topic is structural analysis, and not statistics, very little discussion is given to the actual mechanics of estimation. Do not expect to find computational formulas for various statistical methods. One should consult the standard sources if interested in them. Moreover, it is presumed the reader has some familiarity with the two standard multivariate methods of multiple regression and factor analysis. The emphasis is not on the mechanics of a statistical technique but rather its structural meaning.

I have attempted to present the material in an informal style. Instead of discussing the general case I have chosen small specific examples. Also, to keep the algebra simple, I have not employed matrix algebra. This has made parts of the discussion accessible to the beginner, though it may seem somewhat awkward to the advanced student. I apologize if at times you feel as if you are being talked down to. Please try to sympathize with the concern to write a book that has something to say to both beginner and expert.

I would suggest that you try to read the book with someone else. The corrective nature of dialogue can prevent misapplication better than the addition of another 100 pages. I must warn the reader that a number of sections of the book are not easy to read. I would suggest reading the book with a pencil and calculator in hand. If you take the time to puzzle through various difficult sections, I expect you will find the result to be rewarding.

I have included many examples taken from interesting data sets—for example, the accuracy of psychics, Vietnam protest, concerns of dying cancer patients, and productivity of scientists—which I believe will facilitate the readers' own applications of the methods described in this book. In many cases, however, the models proposed for the data do not fully exploit the structure of the data. The examples then are not meant to demonstrate the successful application of the procedures, but rather the mechanics. Successful applications require full chapters, not two or three pages. Since I provide where possible the full correlation matrix, the reader is invited to fit alternative models.

The text is divided into four major sections, each with three chapters. The first three chapters are introductory—Chapter 1 discussing causal analysis and its role in the social sciences, Chapter 2 presenting a set of simple rules for manipulating covariances, and Chapter 3 introducing path analysis and defining key concepts applied in the remaining chapters.

Chapters 4 to 6 discuss classical econometric methods for estimation of structural parameters. Chapter 4 considers models for which multiple regression analysis estimates causal coefficients; Chapter 5 considers error of measurement in causes and unmeasured third variables; and Chapter 6 briefly introduces feedback models—Chapters 5 and 6 using two-stage least squares to estimate causal parameters.

The third section considers in detail models with unmeasured variables. Chapter 7 discusses models with a single unmeasured variable and multiple indicators, whereas Chapter 8 allows for multiple unmeasured variables; and Chapter 9 details causation between unmeasured variables. All the models discussed in Chapters 7 to 9 are factor analysis models, and maximum likelihood estimation is recommended.

Chapters 10 through 12 consider the application of correlational methods to experimental and quasi-experimental designs. Chapter 10 considers a causal modeling approach to experimentation, as well as additional correlational analyses that could be performed; Chapter 11 considers the analysis of the nonequivalent control group design; and Chapter 12 considers cross-lagged panel correlation analysis. The final chapter of the book, Chapter 13, ties together the loose ends.

A beginning student should read Chapters 2, 3, 4, 7, and the first section of Chapter 10. An advanced student should read the remaining chapters but should concentrate on 5, 8, and 9. I would recommend the following chapters for a course on correlational methods: Chapters 1, 2, 7, 8 (section on the multitrait–multimethod matrix), and 12. For a course on survey analysis, I would recommend Chapters 2 through 9. For a course on experimentation, I would recommend Chapters 2, 3, 4, 7, 8, 9, 10, and 11.

To some purists I may seem to be rather careless. First, I have often unnecessarily limited generalization by employing standardized variables. I did so because I did not want to confuse the beginner. I clearly state the limits of standardization in Chapters 3 and 13. Second, at some points the distinction between population value and estimate is blurred. I did so in order not to have a text filled with distracting symbols. If these practices disturb you, I apologize. But I felt that if I had to sacrifice elegance for the experts in order to obtain clarity for the beginner, I would choose clarity.

A project like this represents not the work of one, but of the community of scholars, and this is especially true of this text. Most of what is contained represents the contributions of others. I would like to thank three persons in particular: First—Don Campbell, who pointed me in this direction and redirected me when I was getting lost; whatever contributions there are in this text are his and not mine. Second—Steven H. Cohen, who suffered with me, pondering many a sentence and analysis; I only wish I were beginning the project now so that he and I could collaborate in it. Third—Pierce Barker, who encouraged and prodded me to finish the project; moreover, he provided detailed feedback for a number of chapters.

Countless others provided helpful feedback throughout the project. Those who occur to me now are Charles Judd, Dean Simonton, Allan Lundy, Jeffrey Berman, William McGarvey, Judith Harackiewicz, Louise Kidder, Lawrence Lavoie, Mary Ellen Kenny, Michael Milburn,

Reid Hastie, James Garvey, and Susan Fiske. Special thanks are due to Alice Mellian and Mary Ellen Kenny, who assisted me in the preparation of the book.

DAVID A. KENNY

Storrs, Connecticut
January 1979

Contents

CORRELATION
AND CAUSALITY

1

Structural Modeling

Given the old saying that "correlation does not imply causation," one might wonder whether the stated project of this book—correlational inference—is at all possible. Correlational inference is indeed possible through the application of standard multivariate statistical methods to a stated structural model. Since a great amount of both confusion and controversy center around the terms "correlational" and "inference," it is first needed to have a common understanding about the meaning of each.

First, *correlational* means a statistical relationship between a set of variables, none of which have been experimentally manipulated. Although correlations and covariances can be computed from experimental data, we usually reserve the term *correlation* for a relationship between unmanipulated variables. Very often random assignment of units to treatment conditions, the backbone of experimental inference, is not possible and we have only correlational data. In such a case we still may wish to make causal inferences.

Second, *inference* means confirmation or disconfirmation of a scientific hypothesis by the use of data. To interpret data, there must be a set of reasonable assumptions about how the data were generated and additional assumptions about how the data can be summarized. The set of assumptions about how the data were generated is usually called the *model,* and data are summarized by *statistical methods.* For at least two reasons, an inference does not imply in any way that the hypothesis is proven true. First, the inference is usually evaluated statistically: At best one knows only the *probability* the results would have obtained by chance given the model. There is no certainty, only a probability. Second, and more important, the model or assumptions on which the inference is based can always be questioned. Every inference but *cogito ergo sum* is based on a set of assumptions. One cannot ascertain simultaneously the validity of both the assumptions and the inference. Modern epistemology tells us that proof is a goal that is never achieved by social scientists or any scientist for that matter. As the ancient Hebrews felt about their God, the scientist should never speak the words truth or proof but always keep them in mind.

Quite clearly the strength of inference is very different for confirmation than for disconfirmation. Especially for correlational inference, disconfirmation is usually more convincing than confirmation. A disconfirmation implies that the data are not compatible with the hypothesis. A confirmation shows the opposite that the data are compatible with the hypothesis. But the data also normally confirm a host of alternative inferences. It shall be seen that the confirmation process can be greatly strengthened by having the hypothesis state not only what should happen with the data but also what should not.

Confirmatory inference is strong then, only if there are no plausible rival explanations of an effect. Campbell and Stanley (1963) in their classic text show how true and quasi-experimental designs may rule out a set of stated plausible rival hypotheses. Although the exact list of rival hypotheses in Campbell and Stanley does not fit very well with correlational designs, the core ideas of that text are extremely helpful in evaluating the strength of correlational inference. For instance, one important idea of Campbell and Stanley is the necessity of tests of statistical significance. Before a relationship can be interpreted, it must be demonstrated that it is not plausibly explained by chance.

The term *correlational inference* should not be taken to mean that various statistics are by themselves inferential. Regression coefficients, factor loadings, and cross-lagged correlations do not, in and of themselves, have an inferential quality. Given a plausible model, a statistic can be used for inferential purposes, but the statistic itself is merely a passive tool. Inference goes on in the head of the researcher, not in the bowels of the computer.

CAUSATION

There is one particular type of inference that we often wish to make from correlational data: a causal inference. A causal statement, to no one's surprise, has two components: a cause and an effect. Three commonly accepted conditions must hold for a scientist to claim that X causes Y:

1. Time precedence.
2. Relationship.
3. Nonspuriousness.

For X to cause Y, X must precede Y in time. Such time precedence means a causal relationship is asymmetric. To see this let X cause Y with a lag in time, and we then have X_t causes Y_{t+k} where the subscript refers to time with $k > 0$. Note that Y_{t+k} cannot cause X_t since this would violate time precedence. (It is true, however, that Y_t could cause X_{t+k}.) Causal relationships are then fundamentally asymmetric while many statistical measures of relationship are symmetric. Implicit in a causal vocabulary is an active, dynamic process which inherently must take place over time.

There is no obvious logical objection to instantaneous causation (although it would be difficult to observe), but philosophers of science have not chosen to assume it. In a sense, something like causation backwards in time happens with rational foresighted creatures. Suppose an individual named John examines the current economic conditions and sees that a depression is inevitable. John then makes a series of economic decisions to cushion himself from the impending effects of the depression (if that is at all possible). One might argue that depression that occurs after John's decisions causes these prior decisions. We can easily see the fallacy in this line of reasoning. It is not the depression per se that causes the decision, but the perception of the impending depression. Like any perception it may not be veridical. If we allowed for the veridical perception of the future (prophecy), it would be empirically impossible to rule out backwards causation.

The second condition for causation is the presence of a functional relationship between cause and effect. Implicit in this condition is the requirement that cause and effect are variables, that is, both take on two or more values. For example, the Surgeon General tells us now that cigarette smoking causes lung cancer. The statement that "smoking causes lung cancer" is actually shorthand for saying that, other things being equal, smoking increases the probability of lung cancer, over not smoking. Thus, to elaborate "smoking causes cancer" we must define two variables: smoking—presence or absence—and lung cancer—again presence or absence. To measure a relationship between two variables, first it is needed to define the meaning of no relationship between variables or, as it is sometimes called, independence. Two variables are *independent* if knowing the value of one variable provides no information about the value of the other variable; more formally, X and Y are independent when the conditional distribution of Y does not vary across X. If variables are not independent, then they are said to be related.

In judging whether two variables are related, it must be determined whether the relationship could be explained by chance. Since naive observers are very poor judges of the presence of relationships, statistical methods are used to both measure and test the existence of relationships. Statistical methods provide a commonly agreed upon procedure of testing whether a sample relationship indicates a relationship in the population.

The third and final condition for a causal relationship is nonspuriousness (Suppes, 1970). For a relationship between X and Y to be *nonspurious,* there must *not* be a Z that causes both X and Y such that the relationship between X and Y vanishes once Z is controlled. A distinction should be made here between a spurious variable and an intervening variable. Variable Z *intervenes* between X and Y if X causes Z and Z in turn causes Y. Controlling for either a spurious variable or an intervening variable makes the relationship between X and Y vanish; but while a spurious variable *explains away* a causal relationship, an intervening variable *elaborates* the causal chain. Many analysts see the issue of spuriousness as the biggest stumbling block in causal analysis, and it has been variously called the third variable problem, the excluded variable, common factoredness, and cosymptomatic relationship. A commonly cited example of spuriousness is the correlation of shoe size with verbal achievement among young children. The relationship is spurious since increasing age causes increasing shoe size and verbal achievement. Spuriousness usually plays a much more subtle role. Much of the discussion of this book centers around the problem of spuriousness.

Besides the three formal requirements of causality time precedence, relationship, and nonspuriousness, there is perhaps a fourth condition for causation which is difficult to state precisely. This fourth condition is that causality implicitly implies an active, almost vitalistic, process. It is difficult to convey this notion of causality formally just as it is to formally define space or time. The philosopher Immanuel Kant has called causality, along with space and time, a *synthetic a priori,* that is, an idea that we bring to our experience of phenomena. This distinctively human bias toward causal explanations has recently become a central topic among social psychologists who study the attribution process.

THE NEED FOR CAUSAL MODELING

Although causal analysis may pervade our everyday life, it is controversial whether causal analysis is an appropriate concern of the social sciences. It is the contention here that causal modeling should have a central position within social research, although there are other very important tasks in the social sciences such as observation, measurement, data reduction, and theory formulation. For at least three reasons causal modeling needs to be applied to social science research:

1. Since most researchers either implicitly or explicitly construct models, a formal development of the method would assist these researchers.
2. Causal modeling can assist the development, modification, and extension of measurement and substantive theory.
3. Causal modeling can give social science a stronger basis for applying theory to solving social problems.

Beginning with the first point, since causal modeling is already being employed, an increased understanding of the process would aid researchers. Most researchers find themselves with large data sets to which they apply statistical methods. Any such method implies a *statistical model*. Common models with which most researchers are familiar are the analysis of variance, multiple regression, and factor analysis models. Very often a statistical model can elegantly and simply summarize the data. Although the fit of the statistical model may satisfy the curiosity of the statistician, it only whets the curiosity of the social scientist since most social scientists gather data to test substantive theory. They invest their egos and reputations in theory, not in statistical models.

Usually the skeleton of a theory is a string of causal statements, although as is later discussed, the heart of any theory is a metaphor or image. Experiments and surveys are conducted to test a set of causal statements which are usually called hypotheses. *The formation of scientific hypotheses is guided by theory and not by a statistical model.* Consider an example. A theory of deviance states that deviance causes social rejection and social rejection in turn causes alienation. It then makes sense to control for social rejection if we are to measure the impact of deviance on alienation. Although it makes theoretical nonsense to control for alienation in estimating the effect of deviance on social rejection, it is perfectly permissible to control for alienation statistically in estimating the relationship between deviance and social rejection. Many of the serious errors in data analysis are due not to lack of knowledge of statistical methods, but to a failure to apply the appropriate statistical method given the conceptual problem. Levin and Marascuilo (1972) have called such errors, Type IV errors.

The ready availability of computer packages contributes to this conceptual slippage between the idea and data analysis. Users can quickly and cheaply apply factor analysis, multiple regression, and analysis of variance to their data. Even though users may understand the statistical assumptions necessary to apply these techniques, they may not fully comprehend the conceptual assumptions. One good example is factor analysis. Many are sophisticated in the ins and outs of rotation, but few have a solid understanding of the relationship of factor analysis to substantive theory. As is seen in Chapter 7, one use of factor analysis is to test for unidimensionality, that is, the set of measures taps a single construct, but few researchers understand how to apply factor analysis to answer this important question. A better understanding of the process of causal modeling would help researchers choose the appropriate statistical method, use it in the correct way, and interpret it intelligently.

The second reason why causal modeling is important is that it can be used to more exactly state theory, to more precisely test theory, and then to more intelligently modify theory. Unfortunately most of what passes for data analysis seems more like a ritual than an investigation into the underlying process of how the data are generated. A researcher who approaches data from a modeling approach is somewhat more likely to learn something new from the data. Ideally the researcher starts with a model or formulates one. Then the researcher determines if the data to be analyzed can estimate the parameters of the model and if the data can falsify the model. Such estimation and testing reveal whether the model is too general, too simple, or just plain wrong. As is seen in Chapter 3, a researcher who carefully follows all the steps of causal modeling is in the position to test theory. Although causal modeling offers an exciting possibility for researchers, it also clearly shows the limits of data analysis in resolving theoretical issues.

A third reason for causal modeling is that it can provide a scientific basis for the application of social science theory to social problems. If one knows that X causes Y, then one knows that if X is manipulated by social policy, *ceteris paribus*, Y should then change. However, if one only knows that X predicts Y, one has no scientific assurance that when X is changed, Y will change. A predictive relationship may often be useful in social policy, but only a causal relationship can be applied scientifically.

Even when a causal law is known, one must take care in applying it. For instance, although the scientific literature is viewed by some as not very supportive of psychotherapy as a cause of "mental health" (see Smith & Glass (1977) for a contrary view), it would be premature to replace it by more "scientific" methods. Often in conventional wisdom resides an implicit knowledge that surpasses the formal knowledge of science. It may be fortunate that political and practical considerations impede the application of social science to social problems. In absence of *strong* theoretical causal models, the practitioners such as social workers and clinicians deserve to be listened to more than the ivory tower academic. It is a sad commentary that the social scientist is often better skilled *not* in designing better social programs to solve social problems but only in evaluating the success or failure of existing social programs (Weiss, 1972).

LIMITS OF CAUSAL MODELING

Although causal modeling is important in the advancement of social science, it has very definite limitations:

1. The research and data must be grounded in a solid foundation of careful observation.
2. The central ideas or driving themes of theory are not usually causal laws but are more likely images, ideas, and structure.
3. Causal modeling is open to abuse.

Regarding the first point, models are built upon qualitative examination of phenomena. Adherents of participant observation often criticize the elaborate "laws" of statistical modelers, and argue that before a tradition can be developed in an empirical domain, a wealth of qualitative lore must be built up. This careful, almost clinical sensitivity to empirical phenomena is not easily taught or appreciated. Sadly, this sensitivity seems to be primarily emphasized in the study of phenomena of behaviors not easily amenable to the quantitative methods, for example, deviance. Careful observational methods form the basis for measurement (institutionalized observation), and measurement in turn is the basis for quantitative data. Let us not forget that the validity of measures in social science is usually nothing more than face validity and face validity is only based on common sense.

Too often, even in the early stages of research on a topic, the researcher is too far separated from the observation of the empirical process. Almost as often the researcher never sees the raw data but only highly aggregated statistical summaries. True enough, interviewing and observation by themselves can lead to mistaken and facile explanations of behavior, but they nonetheless provide a wealthy source of data, albeit difficult to characterize; and from such observations models and hypotheses can be formed.

Second, although theory takes the form of causal statements, the guiding ideas of theory are not those statements but rather an image or an idea. Many of the important ideas of natural science are not causal but are pictures of a process. Although evolution, the periodic table, and the kinetic theory of gases have a mathematical form, they are fundamentally images.

Some causal modelers seem to act as if the major problem facing an area is finding a key cause of the correct variable. For instance, if a researcher is interested in drug use, research will then be defined as measuring all the possible causes of drug use and seeing which is the best predictor. Given finite samples and the multitude of "independent variables," to no great surprise, some variables are found to be significant. Even though most of these effects are usually obvious and unsurprising, elaborate explanations are found for these variables. Then there are a set of marginal but "interesting" variables. How easy it is to delude oneself that one's study is the study to find the key variable. Of course, no other study can then replicate the key variable. Causal modeling provides no certain path to knowledge. In fact, causal models are maximally helpful only when good ideas are tested. Good ideas do not come out of computer packages, but from people's heads.

Third, like any tool causal modeling can be misused. There is no magic in the pages that follow. The techniques to be discussed—factor analysis and multiple regression—are already in common use. This book elaborates how to more carefully and consistently relate these techniques to theory. One cannot take bad data and turn it into gold by calling it a causal model. The potential promise of the method is that it puts the theoretical assumptions up front. Ideally methodological and statistical questions become clarified and disappear, and the focus of the argument is then turned to the substantive assumptions of the researcher.

There will be a temptation to dress up statistical analyses by calling them a "causal analysis." Although the term causal modeling sounds impressive, remember it forces the researcher to make stronger assumptions in order to make stronger conclusions.

CAUSATION AND FREEDOM

There is a final criticism of causal modeling that is put forth. To some, it is repugnant to think of persons being pushed and pulled by a set of causes external to them. Causal models of human behavior give rise to a vision of persons as marionettes pulled by the strings of some set of universal laws. Ironically the same people who object to social science as an enterprise, usually also have an unrealistic view that social science can be used to manipulate human behavior to achieve evil ends. The fear that advertising or behavior modification can change people's lives is an implicit recognition that human behavior is indeed caused. A simple illustration that our behavior is caused is that, even though we are "free" to travel when and where we want, traffic patterns are even more predictable than the weather.

Although we cannot deny that human behavior is caused, our intuitions and traditions are still correct in telling us that people are free. Even highly developed causal models do not explain behavior very well. A good rule of thumb is that one is fooling oneself if more than 50% of the variance is predicted. It might then be argued that the remaining unexplained variance is fundamentally unknowable and unexplainable. *Human freedom may then rest in the error term.* The hard core determinist would counterargue the error is potentially explainable, but at the present moment science lacks an adequate specification of all the relevant causes. Einstein took this approach in his arguments against physical indeterminacy. This, however, is only a conjecture and it seems just as plausible and no less scientific that human behavior is both caused and free. There is no logical or scientific necessity to argue that behavior is totally determined.

Equating unpredictability with freedom may seem peculiar, but Ivan Steiner (1970) in his discussion of perceived freedom defines decision freedom as the condition in which the decision choice is unpredictable. Choices made with no constraints are free. Unpredictability is only part of freedom. The person must also be responsible for the behavior. The velocity of a subatomic particle may be unpredictable, but it is hardly responsible for its position. Human beings are responsible for their behavior.

There may even be a biological basis for human freedom. The fuel of evolution, both biological and cultural, is variation (Campbell, 1974). There can only be evolution if there are a set of different behaviors from which the environment selects. Perhaps nature programmed into the human brain something like a random number generator. This random number generator creates behavioral responses some of which are by chance adaptive. The organism then begins to learn which responses are adaptive to the situation. Culture, socialization, and values continually select from this pool of variable responses. For operant conditioning to work one needs a variable organism to shape. If human beings are fundamentally variable and, therefore, partially unknowable, then traditional social science has erred in presuming that human action is totally determined. Sadly, social science has taken only one track. It has emphasized control and constraint instead of emphasizing freedom and possibility.

Causal modeling can be compatible with the preceding notion of human freedom. It can also allow us to understand the causes of our behavior so that we can *transcend* those causes. This transcendence can be viewed in both a weak or strong sense. The weak sense is that a causal relationship can be mitigated through intervention. For instance, if high employment causes inflation, the government can take steps to reduce inflation when employment is high. If the governmental action has the desired effect of reducing inflation and little or no effect on employment, it would then be possible to reduce or even reverse the relationship between employment and inflation. The *causal* relationship between employment and inflation would still exist, but the statistical relationship between the two variables would vanish. By the stronger sense of transcendence, it is meant that if we understand the rules of the game (i.e., causal relationships), we can decide to change not the variables, but the rules themselves. We may, for instance, decide that rules generated by a free enterprise economy should be abolished and we should play by a set of rules determined by a socialist economic structure. Only by understanding the ways in which we are determined, can we transcend those rules.

CAUSAL LAWS

Now that we have discussed the pluses and minuses of causal modeling, let us begin to discuss causal modeling itself. A causal law is of the general form:

For all Q, X causes Y.

Q refers to the set of objects or persons to whom the causal law applies, X refers to the cause, and Y the effect. Neglected in many causal models is the specification of Q. In social science Q is usually some subset of persons, although Q can also be the set of words, situations, generations, or nations. If Q is the set of persons, usually additional specifications are made. For instance, it is usually assumed that the person is an adult, is awake, is not mentally retarded, and is healthy. Normally these qualifications of the population are implicitly made. Often for a causal law to operate, certain important necessary conditions must hold. For instance, the person must be motivated, must have already learned a relevant behavior, or must attend to various stimuli. In this text the "for all Q" part of a causal statement is usually omitted. One should, however, avoid the reification of causal laws. As Sartre points out (1968, p. 178), causal laws do not operate in heaven but on people:

> About 1949 numerous posters covered the walls in Warsaw: "Tuberculosis slows down production." They were put there as the result of some decision on the part of the government, and this decision originated in a very good intention. But their content shows more clearly than anything else the extent to which man has been eliminated from an anthropology which wants to become pure knowledge. Tuberculosis is an object of a practical Knowledge: the physician learns to know it in order to cure it; the Party determines its important in Poland by statistics. Other mathematical calculations connecting these with production statistics (quantitative variations in production for each industrial group in proportion to the number of cases of tuberculosis) will suffice to obtain a law of the type $y = f(x)$, in which tuberculosis plays the role of independent variable. By this law, the same one which could be read on the propaganda posters, reveals a new and double alienation by totally eliminating the tubercular man, by refusing to him even the elementary role of *mediator* between the disease and the number of manufactured products.

Although it is all too easy to do, let us not lose sight that causal laws in the social sciences refer to people.

Since both the cause X and the effect Y are variables, the relationship between them can be put into some functional form: $Y = f(X)$ called a *structural equation*. The typical functional form is a linear one:

$$Y = b_0 + b_1X$$

The b_1 term is called the *causal parameter* and its interpretation is straightforward. If X is increased by one unit, Y is increased by b_1 units. The relationship is linear because if X is plotted against Y, there would be a straight line with slope b_1 and a Y intercept of b_0.

Instead of just one X causing Y, there may be a set of Xs. These Xs could be combined in a number of ways, but the simplest is an additive combination:

$$Y = b_0 + b_1X_1 + b_2X_2 + b_3X_3 + \cdots + b_nX_n$$

Each causal variable, the Xs, is multiplied by its causal parameter, and these parameter–variable combinations are then summed. One could concoct other mathematical forms, but the linear model is one of the simplest.

Given a set of variables in a causal model, most if not all the information can be summarized in the *covariances* between variables. Statisticians call the covariances the *sufficient* statistics. Before discussing causal modeling any further, let us take a necessary detour and develop the algebra of covariances.

2

Covariance Algebra

The information about the set of parameters of a linear model, causal or otherwise, is all contained in the covariances between variables. To recover the parameters of a model covariances must be manipulated. Thus, to become proficient at causal modeling the researcher must learn how to determine the covariances between variables. A covariance occurs between a pair of variables, say X and Y, and is symbolized here by $C(X,Y)$. The term $C(X,Y)$ is read as the covariance between X and Y. If it is known that $X = aZ + bU$, then it is also known that $C(X,Y) = C(aZ + bU,Y)$ and so within a covariance a variable's equation can be substituted for the variable itself. Covariance is a symmetric measure; that is, $C(X,Y) = C(Y,X)$.

DEFINITION OF COVARIANCE

So far we have only given a symbol for covariance. Mathematically it is defined as

$$E[(X - \mu_X)(Y - \mu_Y)] \qquad [2.1]$$

where E is the expectation or best guess and μ_X is the population mean of variable X and μ_Y of Y. Some texts refer to covariance as σ_{XY}. The unbiased sample estimate of a covariance is given by

$$\frac{\Sigma[(X - \overline{X})(Y - \overline{Y})]}{N - 1} \qquad [2.2]$$

where observations are independently sampled and $\overline{X}$ and $\overline{Y}$ are sample means and N is the sample size. A common computational formula for covariance is $(\Sigma XY - \Sigma X \Sigma Y/N)/(N - 1)$. Covariance is a measure that is in the scale of both variables. If X is measured in feet and Y in yards, their covariance is in feet-yards. If the units of measurement of X or Y are changed, then the value of the covariance changes.

A covariance is a measure of association or relationship between two variables. The covariance is positive if a person who is above the mean on X is expected to be above the mean on Y. With a negative covariance the pattern is reversed: above the mean on X and below the mean on Y or below the mean on X and above the mean on Y. This relationship between X and Y is ordinarily not perfect but only a tendency. Thus if the covariance is positive usually not all the persons above the mean on X are above the mean on Y, but only the "average" person. A zero covariance implies that being above the mean on one variable is in no way predictive of the other variable.

FACTORS THAT INFLUENCE THE SIZE OF A COVARIANCE

Before moving on to the algebra of covariances, consider factors that may make covariance a misleading measure of association: outliers and nonlinearity. As shown by Equation 2.1, a covariance is the sum of cross-product deviations from the mean. Scores that are more deviant from the mean more heavily influence the covariance. In computing covariances researchers should check for outliers that could distort the covariances. An *outlier* is usually said to be three or more standard deviations from the mean. Following Anscombe (1968) there are three different explanations for outliers: (a) measurement errors, (b) multiple populations, and (c) random values.

I suspect that most outliers are due to a malfunction of the measuring instrument or an error in the transfer of data to computer cards. Some of the possible sources of *measurement errors* are that a respondent could misunderstand the instructions, the response could be mistakenly transposed onto coding sheets, the keypuncher could make an error, or the data cards could be out of order. There are indeed all too many sources of error. Errors are often detected by noting whether the obtained value is within the possible range of values. For instance, negative values for either dollars earned and or for reaction time are ordinarily not allowable. For outliers that are errors, the correct value can often be recovered whereas in some cases this is not possible.

A second and more problematic source of outliers is *multiple populations*. The underlying distribution of the variable is bimodal with a second peak that is considerably different from the first. In a study on memory for phone numbers in which one person's performance is abysmally low, the researcher may discover that the person has some organic problem. The researcher should have screened the subjects to exclude from the study those who had organic problems. In a study of income, if one of the sample respondents is a millionaire, this one person could considerably bias the conclusions of the study even if the sample size were large. This person's income may be a social error but it is not a measurement error. Researchers interested in unraveling causal relationships are usually well advised either to exclude from the study any highly deviant group or to oversample them for an additional separate analysis. If the interest is in estimating population parameters, it is unwise to exclude such units.

Finally, outliers are expected by chance. Given a sample size of 2000 and a normal distribution, about 5 persons would fall three or more standard deviations from the mean and about 1 would be four or more. If an outlier cannot be attributed to measurement error or sampling from multiple populations, then it is reasonable to explain the outlier by chance. To reduce the large effect of outliers on covariances, one might *trim* the distribution: Ignore the highest and lowest values or the two highest and lowest. Although trimming is not yet a common practice, statisticians are currently developing this strategy as well as other resistant methods (Tukey, 1977).

Using a covariance as a measure of association presumes that the functional relationship between variables is linear, that is, a straight line functional relationship. For the examples in Figure 2.1, although there is marked functional association between X and Y, their covariance is zero. For all three examples, if the Xs across each Y are averaged, a vertical line is obtained, indicating no linear relationship and zero covariance. (Even though a line is obtained, a perfect vertical or horizontal line indicates zero covariance.) The first example in Figure 2.1 is a curvilinear relationship. A substantive example of a curvilinear relationship is the Yerkes–Dodson law stating that as motivation increases performance increases up to a point, then as motivation continues to increase performance decreases. We all know that moderate amounts of anxiety increase performance, but large doses are debilitating. The second example in Figure 2.1 is a cyclical relationship. A simple substantive example is one in which X is month of the year and Y is mean temperature for the month. Cyclical relationships are rather common when time is one of the variables. The final nonlinear example in Figure 2.1 is an interaction. The relationship between X and Y is linear but the slope varies for two different groups. The slope

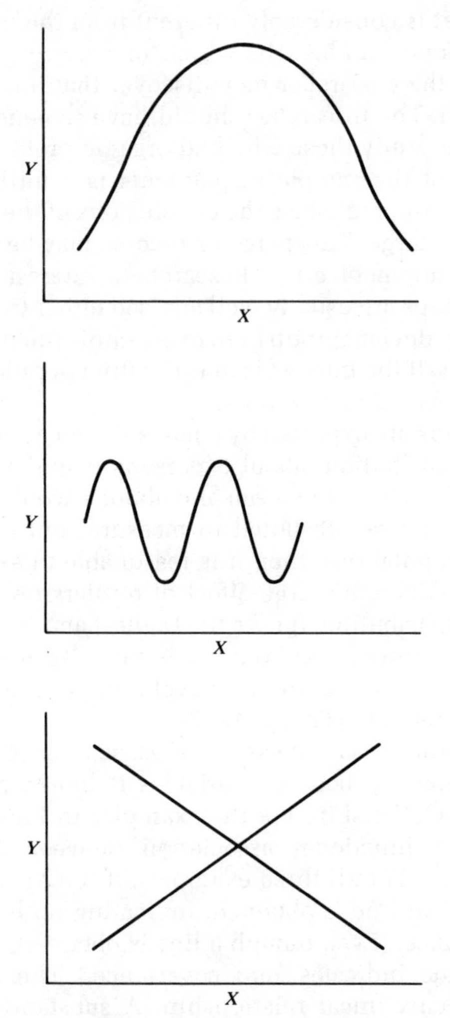

Figure 2.1 Nonlinear relations between X *and* Y*: curvilinear, cyclical, and crossover interaction.*

is positive, or ascending, for one group while the second group has an identical slope of opposite sign. This intersection of lines is sometimes called a crossover, or disordinal interaction. As a substantive example, it seems that among 10-month old infants fear of strangers and cognitive development are positively related (that is, the more fearful are more intelligent) whereas the relationship reverses for 12-month old infants (Jacobson, 1977). Although nonlinearity is a limitation in the generality of covariance as a measure of relationship, it need not be a limitation for the analysis of linear, structural equations. We return to the nonlinear function relationships in Chapter 13 and discuss strategies for coping with the problem.

THE FOUR RULES

So far we have defined a covariance and developed two problems in interpretation: outliers and nonlinearity. We now turn to the central topic of this chapter: covariance algebra. Since all information about causal parameters is contained in the covariances, one must find the key to unlock their precious secrets. The key is contained in four fundamental rules of covariance algebra.

The first rule states that the covariance of a variable with a constant is zero. Thus, if X is a variable and k a constant, $C(X,k) = 0$. Since k is constant across all subjects, its mean ($\overline{k}$) is k and so $k - \overline{k}$ is always zero and every product of $(X - \overline{X})$ is zero. If we plotted X against k, we would obtain a flat line with zero slope indicating no functional relationship between X and k.

The second rule is $C(kX,Y) = kC(X,Y)$. Thus, if a variable is multiplied by a constant, the constant can be factored out of the covariance. Some simple corollaries are $C(kX,bY) = kbC(X,Y)$ and $C(X,-Y) = -C(X,Y)$.

The third rule states that a covariance of a variable with itself, or autocovariance, is simply the variance of the variable. This can be seen by taking the computational formula for covariance and substituting X for Y, yielding

$$\frac{\Sigma\,[(X - \overline{X})(X - \overline{X})]}{N - 1}$$

which is simply the definition of sample variance. Using the second rule, a theorem about variance can be derived. If we multiply a variable by a constant, we multiply the variance of that variable by the constant squared: $C(kX,kX) = k^2C(X,X) = k^2V(X)$ where $V(X)$ denotes the variance of X.

The fourth and final rule is that the covariance of a variable with the sum of two variables is simply the sum of the covariances of the variable with each of the components of the sum: $C(X,Y + Z) = C(X,Y) + C(X,Z)$.

So covariance algebra can be reduced to four simple rules. For

Table 2.1. Rules of Covariance Algebra

Null rule	$C(X,k) = 0$
Constant rule	$C(kX,Y) = kC(X,Y)$
Variance definition	$C(X,X) = V(X)$
Sum rule	$C(X,Y + Z) = C(X,Y) + C(X,Z)$

convenience, they are stated in Table 2.1. To provide names for the rule let us call them the null rule ($C[X,k] = 0$), the constant rule ($C[kX,Y] = kC[X,Y]$), the variance definition ($C[X,X] = V[X]$), and the sum rule ($C[X,Y + Z] = C[X,Y] + C[X,Z]$). It should be noted here one nonrule: $C(X/Y,YZ)$ does not necessarily equal $C(X,Z)$.

APPLICATIONS OF COVARIANCE ALGEBRA

Using the rules one can now expand the following covariance, applying the sum rule first

$$C(aX + bY,X + cZ) = C(aX + bY,X) + C(aX + bY,cZ)$$

again applying the sum rule

$$= C(aX,X) + C(bY,X) + C(aX,cZ) + C(bY,cZ)$$

Now applying the constant rule

$$= aC(X,X) + bC(Y,X) + acC(X,Z) + bcC(Y,Z)$$

and finally the variance definition

$$= aV(X) + bC(Y,X) + acC(X,Z) + bcC(Y,Z)$$

With some familiarity the sum rule need not be successively applied. One can compute the covariance of two sums as the sum of the covariance of all possible cross-products. For instance, find the variance of $X + Y$. The sum of covariances of cross-products is

$$C(X,X) + C(X,Y) + C(Y,X) + C(Y,Y)$$

Adding the two $C(X,Y)$s and using the variance definition yields

$$V(X + Y) = V(X) + V(Y) + 2C(X,Y)$$

The theorem is useful and should be memorized. A more general theorem is that a variance of a sum equals the sum of variances plus twice all possible covariances.

Note that adding a constant does not influence a covariance. First, using the sum rule yields

$$C(X + a,Y) = C(X,Y) + C(a,Y)$$

and the null rule reduces the preceding to $C(X,Y)$.

Many of the familiar measures of linear relationship like regression and correlation can be defined in terms of covariances. For instance, the regression of Y on X (X the predictor and Y the criterion), or b_{YX}, is simply $C(X,Y)/V(X)$. The metric of a regression coefficient is in units of Y per X. As the metric and the formula suggest, the regression coefficient is not symmetric: b_{YX} does not ordinarily equal b_{XY}. Since $b_{XY} = C(X,Y)/V(Y)$, $b_{YX} = b_{XY}$ only if $V(X) = V(Y)$ or $C(X,Y) = 0$. Much more is said about regression coefficients in later chapters, but for the moment let us think of them as the ratio of two covariances.

A scalefree measure of linear relationship is correlation:

$$\frac{C(X,Y)}{[V(X)V(Y)]^{1/2}}$$

or, as it is symbolized, r_{XY} in the sample and ρ_{XY} in the population. As is well known, a correlation varies from $+1$ to -1 and $r_{XY} = r_{YX}$. There is a simple relation between correlation and regression coefficients

$$r_{XY} = b_{YX}\left[\frac{V(X)}{V(Y)}\right]^{1/2}$$

The sign of both a correlation and a regression coefficient depends on the numerator, $C(X,Y)$. Along the same lines, a correlation and a regression coefficient can only be zero if $C(X,Y) = 0$.

A correlation can be thought of as a covariance between two standardized variables. *Standardization* means that the variable has been transformed to have zero mean and unit variance; so to standardize X compute $(X - \overline{X})/V(X)^{1/2}$. An examination of the formula for correlation shows that if X and Y are standardized variables, $C(X,Y) = r_{XY}$ since $[V(X)V(Y)]^{1/2} = 1$ given that X and Y have unit variance through standardization. The fact that the covariance of standard scores is a correlation will be of special help in the development of path analysis in the next chapter.

Using our knowledge of regression coefficients and covariances, one can determine $C(Y - b_{YX}X,X)$:

$$C(Y - b_{YX}X,X) = C(Y,X) - b_{YX}V(X)$$

by the sum and constant rule and variance definition,

$$= C(Y,X) - \left[\frac{C(X,Y)}{V(X)}\right]V(X)$$

by the definition of the regression coefficient,

$$= C(X,Y) - C(X,Y)$$
$$= 0$$

Some readers may recognize that $Y - b_{YX}X$ is the residual from the regression line. It has then just been shown that the residual is uncorrelated with the predictor X.

Covariance algebra is especially useful for showing that what seem to be theoretically meaningful relations between variables are, in fact, only mathematical necessities. For instance, let X be a pretest and Y the posttest. A researcher may wish to correlate "change," $Y - X$, with initial status, X. Let us determine $r_{X\,Y-X}$ in the special case where $V(X) = V(Y)$. Since $C(X,Y - X) = C(X,Y) - V(X)$ and $V(Y - X) = V(Y) + V(X) - 2C(X,Y)$, the definition of correlation implies

$$r_{X\,Y-X} = \frac{C(X,Y) - V(X)}{(V(X)[V(Y) + V(X) - 2C(X,Y)])^{1/2}}$$

and given $V(X) = V(Y)$

$$= \frac{C(X,Y) - V(X)}{(V(X)2[V(X) - C(X,Y)])^{1/2}}$$

and dividing the numerator and denominator by $V(X)$

$$= \frac{r_{XY} - 1}{(2(1 - r_{XY}))^{1/2}}$$

and finally dividing the numerator and denominator by $(1 - r_{XY})^{1/2}$

$$= -\left(\frac{1 - r_{XY}}{2}\right)^{1/2}$$

Thus, the correlation of change with initial status must be negative if the variances of X and Y are equal and r_{XY} is less than one. This fact is part of what is meant by *regression toward the mean*. Note that the difference is negatively correlated with X. Thus those who are high on X have on the average lower standard scores on Y and those who are low on X have higher scores on Y. The interpretation of covariances containing sums and differences must be made with great care.

Another illustration of a potentially misleading covariance is $C(X + Y, X - Y)$ which simply equals $V(X) - V(Y)$. This theorem is useful as a test of the null hypothesis that the variances of two variables measured on the same set of respondents are the same. The null hypothesis of $V(X) = V(Y)$ implies $\rho_{(X+Y)(X-Y)} = 0$, where ρ is the population correlation coefficient. One can then compute $r_{(X+Y)(X-Y)}$ and test whether it significantly differs from zero to test the null hypothesis of the equality of two variances computed from the same set of respondents.

Given a set of variables, their covariances can be summarized by a *covariance matrix*. For instance, the covariance matrix for X, Y, and Z

$$\begin{matrix} C(X,X) & C(Y,X) & C(Z,X) \\ C(X,Y) & C(Y,Y) & C(Z,Y) \\ C(X,Z) & C(Y,Z) & C(Z,Z) \end{matrix}$$

The main descending diagonal of a covariance matrix is the set of autocovariances or variances and the matrix is symmetric with respect to that diagonal. A covariance matrix of standardized variables is a correlation matrix.

Covariance algebra is indeed useful for discovering statistically necessary relationships. However, the main use of it here is with structural equations. We see in the next chapter that if variables are standardized, covariance algebra can be short-circuited by path analysis. But on covariance algebra rests the analysis of linear equations.

3

Principles of Path Analysis

Although most social scientists either explicitly or implicitly use a causal language, they lack a common, systematic approach to causal analysis. Psychologists in particular have relied on a language that grew out of true experimental design. The key notions of independent and dependent variables from experimental design are potentially misleading in a nonexperimental analysis. What follows is an exposition of a set of terms that econometricians and sociologists have found useful in nonexperimental causal inference. Although to some the terms may be foreign and awkward at first, the reader will no doubt find them very useful in conceptualizing the problems in later chapters.

SPECIFICATION

Structural models require a blend of mathematics and theory. Although there are many interesting issues in the mathematics of models, the most difficult questions are those that translate theory into equations. This process of translation is called *specification*. Theory specifies the form of equations. As Blalock (1969) has pointed out, most theories in the social sciences are not strong enough to elaborate the exact form of equations, and so the causal modeler must make a number of theoretical assumptions or specifications. Although every specification can be stated in equation form, the specification should have some justification drawn from theory. As stated in Chapter 1 each is specified as the sum of its causes, an assumption to be returned to in Chapter 13.

However, specifications need not only be based on substantive theory. There are two other sources of specification: measurement theory and experimental design. The theory of test measurement has been well developed, for example, Guilford (1954), and is useful in specifying structural equations. For instance, classical test theory posits that errors of measurement are uncorrelated with the true score. Measurement theory is often helpful in formulating structural models.

Researchers unfamiliar with the logic of experimental design are often unaware that the design of the research can yield additional specifications. For instance, factorial design and random assignment to treatment conditions yield independent causes that are uncorrelated with the unmeasured causes of the effect. Likewise, longitudinal designs may bring with them the specification that certain parameters do not change over time. Successful structural modelers exploit all three types of specifications: substantive theory, measurement theory, and experimental design. Traditionally psychologists focus on experimental design, psychometricians on measurement theory, and econometricians on substantive theory. Rather than choosing one's specification by one's discipline, specifications should be chosen to fit the problem.

What the causal modeler fears more than anything else is *specification error*: One of the assumptions of the model is incorrect. Most models contain misspecification. However, it is not sufficient to criticize a model just because it contains a specification error. One must show that it seriously *biases* estimates of the parameters of the model. It may be that the whole model or parts of it are very robust even with a specification error. A hypothesized specification error must be examined carefully to see exactly how it affects the model.

STRUCTURAL MODELS

Structural models have two basic elements, variables and parameters. Variables, as the name suggests, vary across the persons for whom the causal law applies. One person has an IQ of 100, another of 110, still another of 80. These variables are the raw materials of theories. In this text variables are symbolized by capital letters, for example, I for intelligence and A for achievement. Parameters do not vary across persons, but they describe the whole population. Mean and variance are examples of two parameters. With structural models, parameters are *structural parameters,* and each structural parameter is multiplied by a variable. The sum of these parameter–variable combinations equals the effect variable, and the resultant equation is called a *structural equation*. Structural parameters are symbolized by lower-case letters, $a, b, c,$

. . . , z. A more standard manner of denoting parameters is to use b or β and subscript the cause and effect with the subscript for the effect written first. For instance, if X causes Y, b_{YX} or β_{YX} is the structural parameter. If all the variables are standardized, the structural parameters are called *path coefficients* and ordinarily are designated by lower case ps and subscripted in the same way as the bs and βs. For instance, p_{ji} is the path from i to j where i is the cause and j the effect. I find the subscripting of causal coefficients to be rather clumsy; moreover, the use of the bs implies that the causal parameters are regression coefficients. They are in fact theoretical regression coefficients, but in practice it is often the case that multiple regression analysis does not provide unbiased estimates of causal parameters.

To illustrate a structural equation, income (I) is thought to be the sum of the status of occupation (S) and the amount of education (E); so for person i:

$$I_i = aS_i + bE_i \tag{3.1}$$

All variables are subscripted by person since they vary across persons. Also throughout the text unless otherwise noted all variables are expressed in mean deviation form. For Equation 3.1, I is income minus the group mean. For all future equations the person subscript is dropped and presumed. The lower-case letters in Equation 3.1, a and b, are the causal parameters and their interpretation is straightforward. If S is increased by one unit and if E is kept constant, I is increased by a units. Of course, if a is negative and we increase S by one unit, we decrease I.

Equation 3.1 is an example of a structural equation. By convention the effect is written on the left side of the equation and is sometimes called the left-hand variable; the causes are written on the right side and are sometimes called right-hand variables. A more common vocabulary taken from econometrics is to refer to the causes as *exogenous* variables and to the effects as *endogenous* variables. The causes are sometimes called the independent variables, and the effects are sometimes called the dependent variables. However, this terminology was developed for true experimental research in which the independent variable is always independent because it is experimentally manipulated. In the nonexperimental case, the terms independent and dependent variable are not appropriate. In this text the terms cause or exogenous variable are used and effect or endogenous variable instead of independent and dependent variable.

Structural equations differ from other equations because they represent not only a mathematical relationship, but also a theoretical relationship between cause and effect. For instance, the equation

$$-aS_i = bE_i - I_i$$

is mathematically equivalent to Equation 3.1, but it makes no structural sense.

In structural modeling there is no need to limit ourselves to a single equation, and so there are normally a system of equations. For instance, one might add to Equation 3.1 the following equation:

$$E = cF + dV \qquad [3.2]$$

where E is education, F is father's social class, and V represents all other causes of education. Again all variables are expressed in mean deviation form. Note that E is exogenous in Equation 3.1 and endogenous in 3.2. Variables F, V, and S in these two equations are said to be *purely exogenous* or predetermined variables since they are only causes, while the variable I is only endogenous. A set of structural equations is called the causal or structural model.

There are two types of structural models: hierarchical and nonhierarchical. A nonhierarchical model has feedback loops, either direct or indirect. An example of a direct feedback loop is the case in which X causes Y and Y causes X. An indirect loop occurs when one of the causal links of the feedback loop is indirect, for example, X causes Y and Y causes Z which causes X. As an example of a nonhierarchical model from economics, a model of supply and demand is nonhierarchical since supply causes demand and demand causes supply.

A hierarchical model has no feedback loops. Formally a model is hierarchical if the set of structural equations can be ordered in such a way that an effect in any equation does not appear as a cause in any earlier equation. If the equations cannot be so ordered the model is said to be nonhierarchical. For instance, the following set of equations is nonhierarchical:

$$X = aY + bU$$

$$Y = cZ + dV$$

$$Z = eX + fW$$

since Y causes X and X causes Z which causes Y. The structural model is then nonhierarchical. In this text the central concern is with hierarchical models, although nonhierarchical models are discussed in Chapter 6.

To summarize, a structural model consists of a set of equations. The effect or endogenous variable is on the left side, and on the right side is the sum of the causes or exogenous variables, each causal variable multiplied by a causal parameter. If there are no feedback loops, the model is said to be hierarchical.

PATH ANALYSIS

Often the variables in models have zero mean and unit variance which is called *standardization*; although this greatly simplifies the algebra, it does imply some loss of generality. First, standardization brings about the loss of the original metric which sometimes interferes with interpretation, but in many cases the original metric is arbitrary anyway. Second, occasionally specifications or assumptions are made that involve the original metric. For instance, Wiley and Wiley (1970) suggest that unstandardized error variances may be stable over time. Such models can be handled with standardization but only very awkwardly. Third, the unstandardized metric is more valid when comparing parameters across populations. Fortunately it is generally a simple matter to transform the standardized parameters into unstandardized parameters. The costs of standardization are discussed further in Chapter 13. Unless otherwise stated the reader should presume that all variables are standardized.

Throughout this book a heuristic is used for standardized-hierarchical-linear models: path analysis. Path analysis was first developed by the biologist Sewall Wright (1921) and introduced into the social sciences by the sociologist O.D. Duncan (1966). The working tools of path analysis are the path diagram, the first law, and the tracing rule.

The following set of structural equations can be summarized pictorially in the path diagram in Figure 3.1:

$$X_3 = aX_1 + bX_2$$

$$X_4 = cX_1 + dX_2 + eX_3$$

The rules for translating equations into a path diagram are to draw an arrow from each cause to effect and between two purely exogenous variables draw a curved line with arrowheads at each end. The path diagram contains all the information of a system of equations, but for

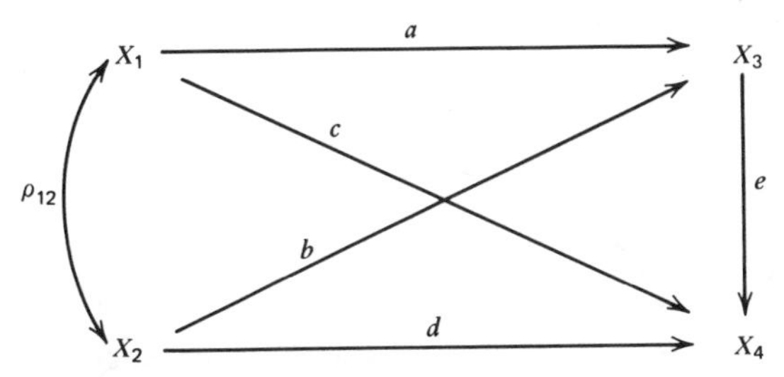

Figure 3.1 Example of a path model.

many models the diagram is easier to comprehend. Most of us feel a little more comfortable examining the picture than the set of equations. Since the diagram contains the set of equations, it is possible to write the equations from the path diagram alone.

In Figure 3.1 the two variables X_1 and X_2 are purely exogenous variables and may be correlated, which is represented by a curved line with arrows at both ends. The correlation between two purely exogenous variables has no causal explanation, that is, the model does not specify what structural relationship brings about the correlation. The researcher may know of no causal model, or may know of many alternative causal models, or may feel the causal model is in the domain of another discipline.

It is rare, indeed, that theory specifies all the causes of a variable, although such a case is discussed in Chapter 9. Therefore, usually another cause must be added that represents all the unspecified causes of the endogenous variable. This residual term is often called the *disturbance*, the term used in this text, or error or stochastic term. The disturbance represents the effect of causes that are not specified. Some of these unknown causes may be potentially specifiable while others may be essentially unknowable. The disturbance is an unmeasured variable and is usually denoted by the letter U, V, or W. The equations for the model in Figure 3.1 adding disturbance terms are as follows:

$$X_3 = aX_1 + bX_2 + fU$$

$$X_4 = cX_1 + dX_2 + eX_3 + gV$$

The path diagram for the model is contained in Figure 3.2. Disturbances can be thought of as purely exogenous variables and so curved lines could have been drawn between U and V and from each to both X_1 and X_2. It has been assumed, however, that the disturbances are uncorrelated with each other and with the purely exogenous variables.

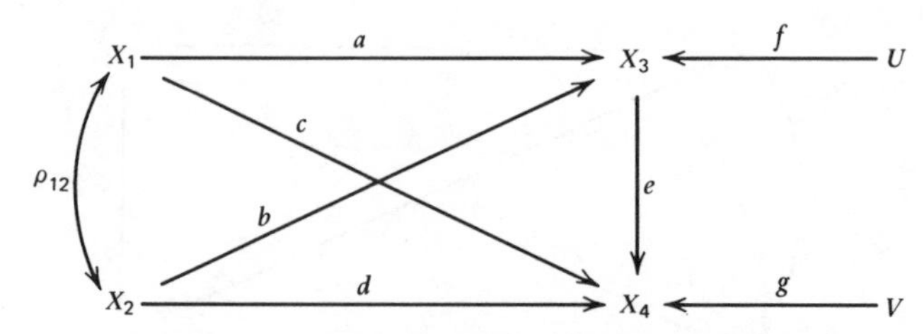

Figure 3.2 Path model with disturbances.

Omitted curved lines imply that the correlation between exogenous variables is zero. The specification of uncorrelated disturbances is one that is often made, and as with any specification it should be theoretically valid and reasonable.

Not all the variables of a structural model are measured. For instance, the disturbance is not measured. However, even some of the other variables may not be measured. For instance, the factor analysis model discussed in Chapter 7 postulates such unmeasured variables.

The First Law

Given a standardized structural model the correlation between any two variables can be derived by the following rule, called the *first law* of path analysis:

$$\rho_{YZ} = \sum_i p_{YX_i}\rho_{X_iZ}$$

where p_{YX_i} is the path or causal parameter from variable X_i to Y, ρ_{X_iZ} is the correlation between X_i and Z, and the set of X_i variables are all the causes of the variable Y. The first law can be shown to follow from covariance algebra. To apply the first law the variable Y must be endogenous. If both variables are purely exogenous, the correlation cannot be broken down and there is no need to apply the first law. To restate the first law verbally: To find the correlation between the variables Y and Z where Y is endogenous, sum the products of each structural parameter for every variable that causes Y with the correlation of each of these variables with the variable Z. A simple procedure to employ is to write all the path coefficients of the endogenous variable including the disturbance. Next to each path write the correlation of the exogenous variable of that path with the variable Z. Multiply each path and correlation and sum the products. For example, the correlation between X_1 and X_3 for the model in Figure 3.2 is

$$\rho_{31} = a\rho_{11} + b\rho_{12} + f\rho_{1U}$$

Since ρ_{11} equals 1 and since no correlation between U and X_1 is assumed (there is no curved line between the two variables), the preceding equation reduces to

$$\rho_{31} = a + b\rho_{12}$$

The reader should work through the remaining correlations for the model in Figure 3.2:

$$\rho_{32} = b + a\rho_{12}$$

$$\rho_{41} = c + d\rho_{12} + e\rho_{13}$$

$$\rho_{42} = d + c\rho_{12} + e\rho_{23}$$

$$\rho_{43} = e + c\rho_{13} + d\rho_{23}$$

$$\rho_{34} = a\rho_{14} + b\rho_{24} + f\rho_{U4}$$

Note that there are two expressions for ρ_{34} since either X_3 or X_4 can be taken as an endogenous variable. The two equations can be shown to be algebraically equivalent. The reader might be puzzled about why $f\rho_{U4}$ is included in the equation for ρ_{34}. Although $\rho_{UV} = 0$, it does not follow that $\rho_{U4} = 0$ but it does follow that $\rho_{3V} = 0$. (This can be seen more clearly once the reader learns the tracing rule.)

The first law can also be applied to the correlation of a variable with its disturbance:

$$\rho_{3U} = a\rho_{1U} + b\rho_{2U} + f\rho_{UU}$$

which reduces to

$$\rho_{3U} = f$$

and similarly

$$\rho_{4U} = g$$

If the disturbance is uncorrelated with the exogenous variables, the path from the disturbance to its endogenous variable equals the correlation of the disturbance with that variable. This equivalence will be useful in later sections.

The Tracing Rule

Path diagrams and the first law hold for nonhierarchical models as well as hierarchical models. But the *tracing rule* does not apply, at least in any simple fashion, to nonhierarchical models. The tracing rule is a simple, nonmathematical rule: The correlation between X_i and X_j equals the sum of the product of all the paths obtained from each of the possible *tracings* between *i* and *j*. The set of *tracings* includes all the possible routes from X_i to X_j given that (a) the same variable is not entered twice and (b) a variable is not entered through an arrowhead and left through an arrowhead. Again using Figure 3.2 one obtains

$$\rho_{31} = a + b\rho_{12}$$

There are two possible tracings: first, a direct path from X_1 to X_3 and a tracing from X_1 to X_2 to X_3. Note that the correlations between purely exogenous variables are treated no differently. A tracing is not allowable from X_1 to X_4 to X_3 since X_4 is entered through an arrowhead and left through an arrowhead. The solution for the correlation of X_1 with X_3 yields an identical expression for both the tracing rule and the first law. They yield identical results only when all the causes of the endogenous variables are purely exogenous. The remaining correlations for Figure 3.2 are

$$\rho_{32} = b + a\rho_{12}$$

$$\rho_{41} = c + d\rho_{12} + ea + eb\rho_{12}$$

$$\rho_{42} = d + ae\rho_{12} + c\rho_{12} + eb$$

$$\rho_{43} = e + ac + ad\rho_{12} + bc\rho_{12} + bd$$

$$\rho_{3U} = f$$

$$\rho_{4V} = g$$

There seem to be different solutions from the tracing rule and the first law for correlations involving X_4. To show the results are the same, though superficially different, examine the solution for ρ_{41} obtained from the first law:

$$\rho_{41} = c + d\rho_{12} + e\rho_{13}$$

Substitute the value obtained from the first law for ρ_{31}

$$\rho_{41} = c + d\rho_{12} + ea + eb\rho_{12}$$

which is identical to the value obtained by the tracing rule. The results obtained by the tracing rule can always be reproduced by using the results from the first law. In general, the first law should be preferred since it ordinarily yields a result that is simpler than the tracing rule. Moreover, there is more chance of making an error with the tracing rule since one may not be sure if all possible tracings have been made. The tracing rule is nonetheless useful, particularly with models with unobserved variables.

It may be helpful to work through another example. A researcher is interested in the causal model for verbal (X_4) and mathematical achievement (X_5). Theory tells her that parental socioeconomic status (X_1), intelligence (X_2), and achievement motivation (X_3) cause the two achievement variables. She also assumes that there are other common causes of the two achievement variables. The structural equations are:

$$X_4 = aX_1 + bX_2 + cX_3 + dU$$

$$X_5 = eX_1 + fX_2 + gX_3 + hV$$

with the side conditions

$$\rho_{1U} = \rho_{2U} = \rho_{3U} = \rho_{1V} = \rho_{2V} = \rho_{3V} = 0$$

$$\rho_{UV} \neq 0$$

The assumption of omitted causes of X_4 and X_5 is embodied by the correlation of the disturbances. The path diagram is in Figure 3.3, and using either the first law or tracing rule the correlations of the variables are:

$$\rho_{41} = a + b\rho_{12} + c\rho_{13}$$

$$\rho_{42} = a\rho_{12} + b + c\rho_{23}$$

$$\rho_{43} = a\rho_{13} + b\rho_{23} + c$$

$$\rho_{51} = e + f\rho_{12} + g\rho_{13}$$

$$\rho_{52} = e\rho_{12} + f + g\rho_{23}$$

$$\rho_{53} = e\rho_{13} + f\rho_{23} + g$$

$$\rho_{4U} = d$$

$$\rho_{5V} = h$$

$$\rho_{4V} = d\rho_{UV}$$

$$\rho_{5U} = h\rho_{UV}$$

Using the first law either

$$\rho_{54} = e\rho_{14} + f\rho_{24} + g\rho_{34} + h\rho_{V4}$$

or equivalently

$$\rho_{45} = a\rho_{15} + b\rho_{25} + c\rho_{35} + d\rho_{U5}$$

and the tracing rule

$$\rho_{45} = ae + bf + cg + af\rho_{12} + be\rho_{12} + ag\rho_{13} + ec\rho_{13} + bg\rho_{23} + fc\rho_{23} + dh\rho_{UV}$$

As can be seen for ρ_{45}, the expressions can be complicated, but if the simple rules are followed it all becomes routine.

It should be noted that the laws of path analysis hold for all correlations in a causal model, including the correlation of a variable with itself. Take, for instance, the following equation:

$$X_3 = aX_1 + bX_2 + cU$$

where

$$\rho_{1U} = \rho_{2U} = 0$$

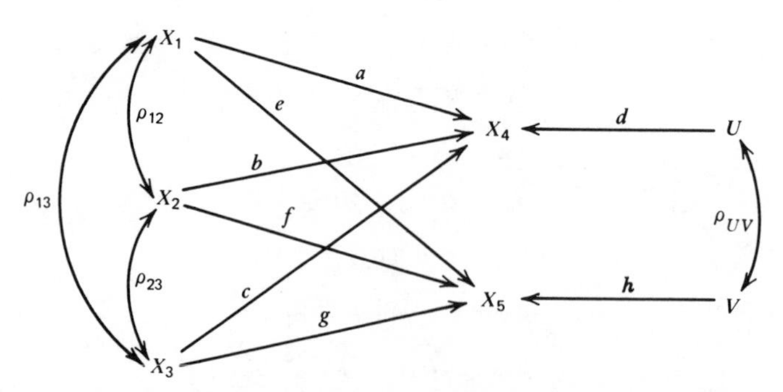

Figure 3.3 Model with correlated disturbances.

Using the first law

$$\rho_{33} = a\rho_{13} + b\rho_{23} + c\rho_{3U}$$

Since we know that ρ_{33} must equal 1, it follows that the preceding equation equals 1.

To use the tracing rule to find ρ_{33} simply draw another X_3 as was done in Figure 3.4. Now find all the possible tracings yielding

$$\rho_{33} = a^2 + b^2 + 2ab\rho_{12} + c^2 \qquad [3.3]$$

A third solution for ρ_{33} is to solve for the variance of $aX_1 + bX_2 + cU$ which is known to be one because of standardization. This would be an expression identical to Equation 3.3, a fact that is later useful in estimating the path from the disturbance to the endogenous variable.

The reader now knows how to solve for correlations of a given standardized, hierarchical, linear model. A major goal of structural modeling is not to find such correlations, but rather to find the values of the causal parameters. The path coefficients of a given model are solved for by a four-stage process:

1. Measure as many variables of the model as possible.
2. Compute the correlations between the measured variables.
3. Using the first law or the tracing rule, derive the formulas for correlations.
4. Substituting in the correlations computed from data, solve for the path coefficients.

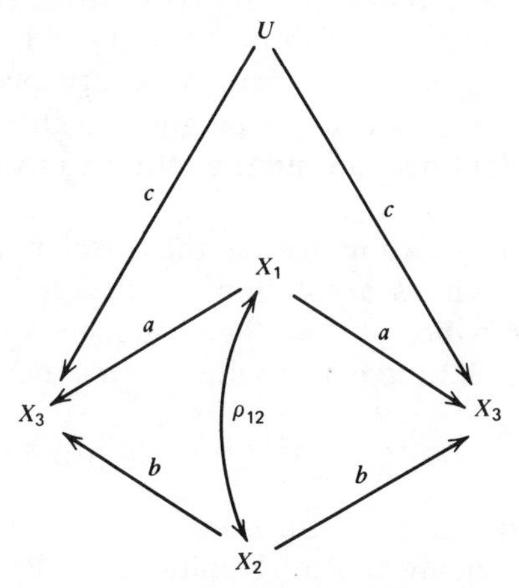

***Figure 3.4** Path diagram with the effect written twice.*

The first two steps are simple enough for anyone who has gathered data previously. The reader has just learned how to do the third step, and with some practice the reader will find that it is simple to do. The trouble comes with the last step; as is seen in the next section it may not even be possible to solve for the parameters from the correlations.

IDENTIFICATION

Determining whether estimation of parameters of the model is possible brings one to the issue of *identification*. Although identification may seem to be a strange term to use when speaking of a set of equations, it is the term that econometricians have chosen. Identification has nothing to do with the number of observations (there must be at least as many observations as parameters plus two), but rather with the number of correlations between the measured variables. A *necessary* but not sufficient condition to be able to identify and, therefore, estimate the causal parameters of a set of structural equations is that the number of correlations between the measured variables be greater than or equal to the number of causal parameters. This necessary condition is called the *minimum condition* of identification. Given that the two are equal, it *may* be possible that the set of causal parameters is *just-identified*; that is, there is one and only one estimate for each causal parameter. If there are more correlations than parameters, the structural model is said to be *overidentified*; that is, there is more than one way of estimating a causal parameter in the system. For instance, the logic of the F test in one-way analysis of variance is that there are two estimates of error variance under the null hypothesis: variability within and between groups. The error variance is then overidentified. A set of equations is said to be *underidentified* if there are more parameters than correlations. If a parameter is not identified, an infinite number of values would satisfy the equations.

As was stated earlier, knowing that the correlation of a variable with itself equals unity yields a solution for the path coefficient of the disturbance if the values of the other parameters are known. For instance, in Equation 3.3 the path coefficient for the disturbance equals

$$c^2 = 1 - a^2 - b^2 - 2ab\rho_{12} \qquad [3.4]$$

Given that the metric of the disturbance is arbitrary, by convention one takes the positive square root of Equation 3.4. Parameter c is a *constrained* parameter; that is, its value is determined by the other parameters of the model. The implication of this for the identification issue is that one need not worry about the identification of the paths from the disturbances to endogenous variables, but one need consider only the identification of the other parameters of the model. When the disturbance is uncorrelated with the exogenous variables and given that

$$Y = \sum_i p_{Yi} X_i + p_{YU} U$$

the solution for path from U to Y is

$$p_{YU} = (1 - \sum_i p_{Yi} \rho_{iY})^{1/2}$$

Procedurally, how does one assess the status of identification? First one determines the number of correlations between the observed or measured variables. If n variables are measured, the number of correlations is $n(n - 1)/2$. Then count the number of parameters making sure to include (a) all the path coefficients, (b) all correlations between purely exogenous variables, (c) all correlations between disturbances but not to include the path coefficients of the disturbances. For instance, for the model in Figure 3.3, if X_1 through X_5 are measured, there are 10 correlations ($5 \times 4/2 = 10$) and 10 causal parameters: six path coefficients, a, b, c, g, e, and f; three correlations between purely exogenous variables, ρ_{12}, ρ_{13}, and ρ_{23}; and one correlation between disturbances, ρ_{UV}. The model may be just-identified (it is, in fact). I say *may* because one parameter might be overidentified and all the remaining parameters might be underidentified. Remember, having the number of correlations equal to the number of parameters is only a minimum condition for identification.

The model in Figure 3.3 would be underidentified if X_3 were not measured since there still are 10 parameters but only six correlations. The model would be overidentified if it is assumed that the disturbances were uncorrelated since there would be nine parameters and 10 correlations.

If a model is overidentified, there are two estimates of a causal parameter or some function of causal parameters; that is, one can express a causal parameter as two or more different functions of the correlations. If the functions of correlations are set equal to each other, there is an equation that says two sets of correlations equal each other. The resultant equality is called an *overidentifying restriction* on the model. Often a subset of overidentifying restrictions implies one or more of the other overidentifying restrictions. The smallest set of restrictions that implies the remaining ones yields the number of independent restrictions on the model. If a model is truly overidentified, the number of free parameters plus the number of independent overidentifying restrictions is equal to the number of correlations. Overidentifying restrictions can be viewed as a constraint on the structure of a correlation matrix. Thus, the number of free correlations is effectively reduced by the number of overidentifying restrictions. One must then subtract the number of independent restrictions from the number of correlations to determine if the minimum condition is met. These restrictions play an important role in structural modeling because they can be used to test the validity of the model.

Any overidentifying restriction can be stated as a null hypothesis and then tested. Rarely will an overidentifying restriction perfectly hold in the sample, but if the model is valid it should hold within the limits of sampling error. Overidentifying restrictions also increase the efficiency of parameter estimation (Goldberger, 1973). If there are two estimates of the same parameter, those estimates can be pooled to obtain a new estimate whose variance is less than or equal to the variance of either original parameter estimate.

A Just-Identified Model

Before proceeding any further with the concept of identification, consider an example. The equation for the path diagram in Figure 3.5 is

$$X_3 = aX_1 + bX_2 + cU$$

The correlations are

$$\rho_{31} = a + b\rho_{12} \quad [3.5]$$

$$\rho_{32} = b + a\rho_{12} \quad [3.6]$$

Adding ρ_{12} there are three correlations $(3 \times 2/2)$ and three parameters,

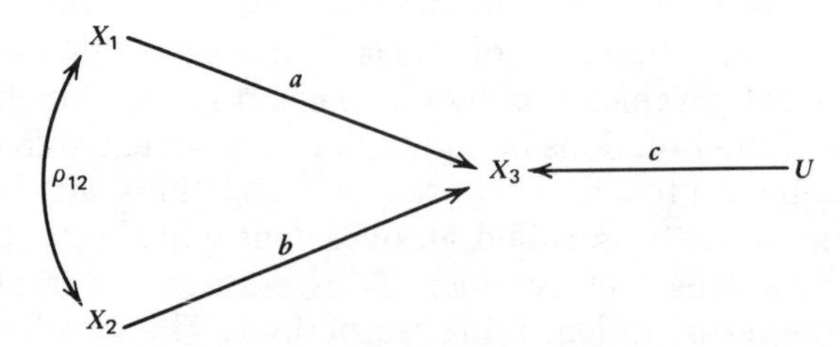

Figure 3.5 Model with observed variables as causes.

a, b, and ρ_{12}, and the model is possibly just-identified. Solutions for a and b exist since there are two linear equations with two unknowns. Multiplying ρ_{12} times Equation 3.6 yields

$$\rho_{23}\rho_{12} = b\rho_{12} + a\rho_{12}^2 \quad [3.7]$$

If we subtract Equation 3.7 from Equation 3.5 we have

$$\rho_{13} - \rho_{12}\rho_{23} = a - a\rho_{12}^2$$

We now solve for a yielding

$$a = \frac{\rho_{13} - \rho_{12}\rho_{23}}{1 - \rho_{12}^2} \quad [3.8]$$

And by analogy

$$b = \frac{\rho_{23} - \rho_{12}\rho_{13}}{1 - \rho_{12}^2} \quad [3.9]$$

As was seen in Equation 3.4, $c^2 = 1 - a^2 - b^2 - 2ab\rho_{12}$. By knowing the model was just-identified one knows, at least, that it may be possible to find solutions for a, b, and c.

The reader may have noticed something interesting about Equations 3.8 and 3.9. The solution for parameter a is simply the beta weight or standardized regression coefficient of X_3 on X_1 controlling for X_2, whereas b is the beta weight of X_3 on X_2 controlling for X_1. Compare Equations 3.8 and 3.9 with p. 192 in McNemar (1969). The reader may breathe a sigh of relief since one will not need to solve a different set of simultaneous equations for every causal model. As shall be seen, regression coefficients, partial correlations, factor loadings, and canonical coefficients estimate causal parameters of different models. After progressing through the book, the reader should be able to tell what statistical technique estimates the parameters for a given path model, which should give the reader more insight to the causal interpretation of these statistical techniques.

An Overidentified Model

For the model in Figure 3.6 variables X_1 through X_4 are measured while the disturbances and F are unmeasured. The disturbances of X_1 are correlated with X_2 and those of X_3 with X_4. (As an exercise write out the

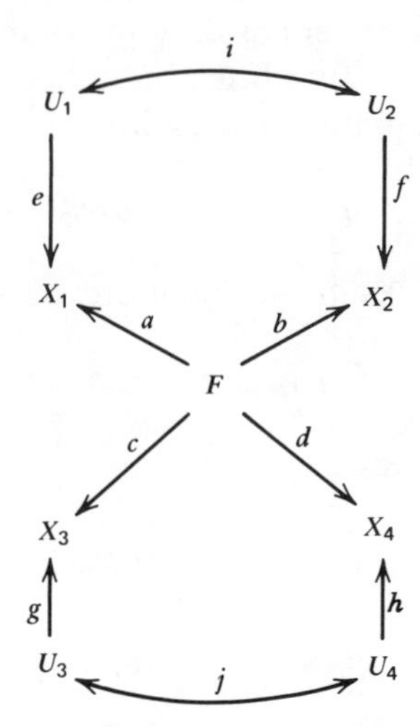

Figure 3.6 Model with an overidentifying restriction.

structural equations for the model in Figure 3.6.) There are six correlations (4 × 3/2) and six free parameters: four path coefficients, a, b, c, and d; and two correlations between disturbances, i and j. The model is possibly just-identified. The correlations obtained by the tracing rule are

$$\rho_{12} = ab + efi$$

$$\rho_{13} = ac$$

$$\rho_{14} = ad$$

$$\rho_{23} = bc$$

$$\rho_{24} = bd$$

$$\rho_{34} = cd + ghj$$

Try as you can, you will never be able to express the parameters in terms of the correlations. Even though there are six correlations and six free parameters satisfying the minimum condition of identifiability, none of the parameters are identified. This is due to the overidentifying restriction of

$$\rho_{13}\rho_{24} = \rho_{14}\rho_{23}$$

since both $\rho_{13}\rho_{24}$ and $\rho_{14}\rho_{23}$ equal $abcd$. Recall that an overidentifying restriction is a constraint on the correlation (covariance) matrix of the measured variables. The overidentifying restriction, even if true, should only perfectly hold in the population. Given the overidentifying restriction plus the six parameters there is not enough information to estimate the parameters of the model, since the overidentifying restriction reduces the number of free correlations to five, which is less than the number of free parameters. Even though none of the parameters of the model are estimated, the model can still be tested by taking the overidentifying restriction as a null hypothesis. Such is also the case for cross-lagged panel correlation, which is discussed in Chapter 12.

Underidentification

Should one give up if the model is underidentified? As has been already stated, even though as a whole the model may be underidentified (a) various parameters of the model may still be just-identified, (b) some parameter may even be overidentified, or (c) as in Figure 3.6 no parameters are identified, but there may be an overidentifying restriction on the model. But, alas, in many cases none of the preceding three conditions hold. To achieve identification or even overidentification various strategies can be employed. One strategy is to measure more variables. This is useful when an exogenous variable is correlated with the disturbance or when an exogenous variable is measured with error. Unfortunately, adding more variables and more equations can at times only make matters worse. Another strategy to bring about identification is to reduce the number of parameters. There are three common means to do this. First, one can assume that certain parameters are zero. As we saw and shall continue to see, one often assumes that the disturbance is uncorrelated with the exogenous variables. One may assume that certain paths are zero as in Chapter 5 with the method that econometricians call "instrumental variable" estimation. The second method to reduce the number of parameters is to set two of them equal to each other, thereby decreasing the number of free parameters by one. For instance path coefficients are set equal as in the case of cross-lagged panel correlation in Chapter 12, or in the case of disturbances set equal across equations, or correlations between disturbances are set equal for the multitrait–multimethod matrix in Chapter 8. The third and last strategy is to make a proportionality constraint. Given four parameters, a through d, one might assume $a/b = c/d$. Thus only three parameters are free since if three are known the fourth can be solved for. For instance, if a, b, and c are known, d equals bc/a. There are other procedures available for reducing the number of parameters but the three preceding are most common.

All these strategies are viable only if they can be justified by substantive theory. Making new specifications just to be able to identify the parameters of a causal model is perhaps the worst sin of causal modelers. Obviously, identification is necessary for causal modeling, but one must not sloppily add constraints just to achieve identification.

Duncan (1976) has suggested yet another approach to underidentification. He suggests fixing one parameter, or as many as needed, to a reasonable value and solving for the remaining parameters. Then by changing the original parameter again, another set of parameter estimates are obtained. This is done repeatedly, until a range of possible values are obtained. Often it will happen that the obtained range of values is, under this procedure, rather limited. Occasionally, there are no reasonable solutions; for instance, a correlation between disturbances is larger than one or a parameter's solution is imaginary (the square root of a negative value). The lack of a reasonable solution would indicate some sort of specification error.

Empirical Underidentification

Even though a system of equations may be identified in principle, in practice there may be no solution for a parameter. It may happen that when the correlations or covariances are substituted into the expression that should estimate the parameter, the denominator of that expression is zero or very near zero. Since division by zero is algebraically undefined, there is no solution. The case in which the denominator of an expression that estimates a parameter is equal or nearly equal to zero, is called in this text *empirical underidentification*. As is seen in the next chapter multicollinearity is an example of empirical underidentification.

After one has determined that a set of equations is identified in principle and obtained estimates of each parameter, one should carefully examine the denominator of each estimate and note the condition under which it equals zero; this condition defines empirical underidentification.

Although the estimate is defined when the denominator is nonzero, if the denominator is very close to zero the estimate is practically useless since its standard error may be huge. Thus empirical underidentification is defined both by zero or near zero denominators. Of course, there is the problem of how to define "near zero." When working from a correlation matrix, I have found that if the denominator is less than .1, then the estimate is often so unstable as to be worthless.

A Single Unmeasured Variable

One need not measure all the variables in a model besides disturbances to achieve identification. Consider the following model

$$X_1 = aF + eU_1$$
$$X_2 = bF + fU_2$$
$$X_3 = cF + gU_3$$

with the following side conditions

$$\rho_{U_iF} = 0 \ (i = 1,2,3)$$
$$\rho_{U_iU_j} = 0 \ (i \neq j)$$

(As an exercise draw a path diagram for the preceding set of equations.) The model is just-identified with X_1, X_2, and X_3 measured and F being unmeasured. There are three correlations and three free parameters, a, b, and c. The correlations are

$$\rho_{12} = ab$$
$$\rho_{13} = ac$$
$$\rho_{23} = bc$$

The parameters are then

$$a^2 = \frac{\rho_{12}\rho_{13}}{\rho_{23}}$$

$$b^2 = \frac{\rho_{12}\rho_{23}}{\rho_{13}}$$

$$c^2 = \frac{\rho_{13}\rho_{23}}{\rho_{12}}$$

This is a single factor model where a, b, and c are factor loadings (cf. Duncan, 1972; Harmon, 1967).

STATISTICAL ESTIMATION AND INFERENCE

The discussion has been mainly restricted to the population by assuming that the population values of correlation coefficients are known. Rarely if ever are these correlations known and in practice the researcher must use sample estimates. Given only sample data, statistical methods must be used to estimate parameters. Ideally, the estimates should be sufficient, unbiased, consistent, and efficient. (See Hays (1963) for definitions.) Fortunately, most multivariate statistical methods can be adapted to estimation of the parameters of structural models.

If the equations are simple enough or the researcher ingenious enough, it may be possible to solve for the structural parameters from the set of equations that are obtained from the tracing rule or the first law. Such a solution is commonly called a *path analytic solution*. Often one can just substitute the sample estimates of correlations into the path analytic solution values. This ordinarily gives good estimates if the model is just-identified. However, if the parameter is overidentified there will be multiple path analytic solutions. For such cases traditional statistical methods are to be preferred.

Statistical methods are also needed to find standard errors of parameter estimates, and to test whether a given parameter is zero or whether two parameters are equal. Statistical methods are, in addition, used in testing overidentifying restrictions. To make these significance tests, ordinarily additional specifications must be made which usually concern the distribution of a variable or the disturbance.

REDUCED FORM

With some models it is simpler not to deal with the parameters themselves but to work with the *reduced form* coefficients. To solve for the reduced form coefficients the structural equations are expressed solely in terms of the purely exogenous variables. As an example, given the following set of equations

$$X_2 = aX_1 + bU \qquad [3.10]$$

$$X_3 = cX_1 + dX_2 + eV \qquad [3.11]$$

within Equation 3.11 one can substitute for X_2 what it equals in Equation 3.10 or

$$X_3 = cX_1 + d(aX_1 + bU) + eV$$

$$X_3 = (c + da)X_1 + dbU + eV$$

The reduced form coefficient for X_1 is not a single path coefficient but a function of coefficients $(c + da)$. At times it is easier to solve for the reduced form coefficients and then solve for the path coefficients. Furthermore, as is seen in the next chapter, reduced form is useful for solving for indirect paths from exogenous variables to an endogenous variable (Alwin & Hauser, 1975; Finney, 1972).

TINKERING WITH EQUATIONS

Sometimes for social policy considerations the researcher wants to know the effect of altering structural equations. For instance, this was a central concern of Jencks et al. (1972) in the *Inequality* volume. Jencks et al. continually ask questions such as, "If the father's occupational status no longer affected the amount of the child's education, by how much would the variance of child's education be reduced?", or, "Would reducing inequality in one domain significantly reduce inequality in another domain?" For a structural equation many parameters affect the variance of the endogenous variable. Given the *unstandardized* structural equation of

$$X_3 = aX_1 + bX_2$$

its variance is defined as

$$V(X_3) = a^2V(X_1) + b^2V(X_2) + 2abC(X_1, X_2) \qquad [3.12]$$

using the variance definition and sum rule of the previous chapter. Examining Equation 3.12 note that there are three ways to affect variance of the endogenous variable X_3:

1. Change the causal parameters a and b.
2. Change the variance of the exogenous variables.
3. Change the covariance between exogenous variables.

Mathematically, one is not free to alter these parameters in any way since a number of side conditions must be met. For instance, the absolute value of any correlation must be less than or equal to one, as must the absolute values of the partial correlations of $\rho_{12.3}$, $\rho_{13.2}$, and $\rho_{23.1}$.

If the variance of X_1 is altered, the covariance between X_1 and X_2 may as well be affected. If it is assumed that the change in variance equally affects the component correlated with X_2 and the component uncorrelated, then the altered covariance is equal to the original covariance times the square root of the factor by which the variance was changed.

If there is a *set* of structural equations, special care must be taken when a variable is altered that is endogenous in one equation and exogenous in another equation. If its variance is changed one must specify exactly how the structural equation of the endogenous variable is altered. Different changes in its equation, even if the resultant variances are the same, have different consequences for the variances of the other endogenous variables.

Finally, one should not assume that, by reducing the variance of an exogenous variable, one automatically decreases the variance of the endogenous variable. Though rather uncommon, it is possible for the variance to increase. Similarly an increase in the variance of an exogenous variable may result in a decrease in the variance of the endogenous variable.

CONCLUSION

A great amount of material has been covered. The reader should know how to express correlations in terms of causal parameters and how to draw a path diagram from a set of equations and vice versa. The reader should also know what is meant by the following terms: endogenous and exogenous variable, standardized–hierarchical–linear structural model, specification, identification, specification error, overidentifying restriction, and reduced form coefficients. The remainder of the book is only the application of these concepts to different structural models.

The steps of structural modeling are:

1. From theory draw up a set of structural equations.
2. Choose a measurement and design model.
3. Respecify the structural model to conform to design and measurement specifications.
4. Check the status of identification of the model.
5. Estimate the correlations (covariances) between the measured variables and from the correlations (covariances) estimate the parameters of the model and test any overidentifying restrictions.

In the next chapter it is shown that multiple regression can be used to estimate the causal parameters when measured variables cause measured variables.

4

Models with Observed Variables as Causes

To many readers multiple regression is synonymous with path analysis. Although it is true that most empirical applications of path analysis use multiple regression analysis to estimate causal parameters, many models require the use of more sophisticated techniques to estimate their parameters. This chapter elaborates the assumptions, or specifications, that must be made to interpret ordinary multiple regression coefficients as path coefficients. More complicated techniques are discussed in later chapters. The chapter is divided into three sections. The first section briefly explains the elements of multiple regression analysis. The second section details how regression coefficients can be interpreted as causal parameters. The third section considers a number of technical issues.

MULTIPLE REGRESSION

Regression analysis begins with a set of predictor variables, say father's occupation (F) and intelligence (I), and a criterion, say school grades (G). What the researcher is interested in is the "best" linear combination of the predictor variables that predicts the criterion. For the grade example such a linear prediction equation is

$$\hat{G} = b_1F + b_2I$$

where $\hat{G}$ is predicted grades. Given errors of prediction,

$$G = b_1F + b_2I + E \qquad [4.1]$$

where E is lack of fit, or residual term. All variables but E are standardized. At issue in predicting G is the choice of the coefficients b_1 and b_2, or, as they are commonly called, regression coefficients. The coefficients are chosen to maximize the correlation of the quantity $b_1F + b_2I$ with G. It can be shown that maximizing this correlation is the same as minimizing the sum of squared residuals. Minimizing sums of squares is accomplished by the method of least squares, from which it follows that the correlation of any predictor variable with the residual term must be zero. Using the first law to correlate each of the predictors with grades, the resulting equations are:

$$r_{GF} = b_1 + b_2 r_{FI}$$

and

$$r_{GI} = b_1 r_{FI} + b_2$$

Since there are two linear equations in two unknowns, the unknowns b_1 and b_2 can be solved for:

$$b_1 = \frac{r_{GF} - r_{FI} r_{GI}}{1 - r_{FI}^2} \qquad [4.2]$$

and

$$b_2 = \frac{r_{GI} - r_{FI} r_{GF}}{1 - r_{FI}^2} \qquad [4.3]$$

(Compare with Equations 3.8 and 3.9 of the previous chapter.) Note that each regression coefficient is simply a function of both the correlations among the predictor variables and the correlations of the predictors with the criterion. As the number of predictor variables increases, the equations can be solved in the same way, although not surprisingly the regression coefficients are equal to a more complicated function of the correlations. Fortunately, virtually every computer center has a multiple regression program.

Kerchoff (1974, p. 46) reports the correlations between a number of variables for 767 twelfth-grade males. In Table 4.1 some of these correlations are reproduced. Of particular interest now are the correlations between F, I, and G:

$$r_{FI} = .250$$

$$r_{GF} = .248$$

Table 4.1. Correlations Taken from Kerchoff[a,b]

I	1.000						
S	−.100	1.000					
E	.277	−.152	1.000				
F	.250	−.108	.611	1.000			
G	.572	−.105	.294	.248	1.000		
X	.489	−.213	.446	.410	.597	1.000	
A	.335	−.153	.303	.331	.478	.651	1.000
	I	S	E	F	G	X	A

[a] I, intelligence; S, number of siblings; E, father's education; F, father's occupation; G, grades; X, educational expectation; A, occupational aspiration.
[b] N = 767.

and

$$r_{GI} = .572$$

Solving for b_1 and b_2 by Equations 4.2 and 4.3, respectively, yields

$$b_1 = \frac{.248 - (.250)(.572)}{1 - .250^2} = .112$$

$$b_2 = \frac{.572 - (.250)(.248)}{1 - .250^2} = .544$$

The interpretation of the regression weights in the standardized case or, as they are more commonly called, *beta weights*, is straightforward. If someone is one standard deviation above the mean in father's occupation and at the mean in intelligence, he is on the average .112 of a standard deviation above the mean in school grades. Similarly, someone at the mean in father's occupation and one standard deviation above in intelligence is .544 of a standard deviation above the mean in school grades.

Often a researcher is interested in how well the regression equation predicts the criterion. One measure is to obtain the variance of the predicted scores. The variance of the predicted score is defined as

$$V(\hat{G}) = C(b_1F + b_2I, b_1F + b_2I)$$
$$= C(b_1F + b_2I, G - E)$$

Since $C(F,E) = C(I,E) = 0$ given least squares, it follows that

$$V(\hat{G}) = b_1 r_{GF} + b_2 r_{GI} \qquad [4.4]$$

which holds for the standardized case. Another measure of fit is the multiple correlation which is the correlation of the predicted score $\hat{G}$, or $b_1F + b_2I$, with G. The multiple correlation, or $R_{G(FI)}$, is

$$(b_1 r_{GF} + b_2 r_{GI})^{1/2}$$

since $C(G, b_1F + b_2I) = b_1 r_{GF} + b_2 r_{GI}$, $V(G) = 1$ and $V(b_1F + b_2I)$ equals Equation 4.4. Note that the variance of the predicted scores, Equation 4.4, is simply the multiple correlation squared. Solving for the multiple correlation of our example yields

$$[(.112)(.248) + (.544)(.572)]^{1/2} = .582$$

The variance of the residuals is simply $1 - R_{G(FI)}{}^2$ since G is standardized and the predicted G is uncorrelated with the residuals.

In general, the multiple correlation is the square root of the sum of the products of each regression coefficient and the corresponding correlation of the predictor with the criterion. The variance of the errors is simply one minus the multiple correlation squared.

Very often researchers are interested in more than predicting the criterion: They want the "best" statistical estimates of the betas and they want to test hypotheses about the regression equation. For estimates of the betas to be unbiased, consistent, and efficient, the following must be assumed:

a. Independence: Each observation should be sampled independently from a population. Independence means that the errors, the Es, are statistically independent; that is, the covariances of errors are zero. The sampling unit should be the same as the unit in the analysis; otherwise this assumption will be violated. Independence is also violated if the data are repeated measurements on a single unit since scores tend to be *proximally autocorrelated*, that is, data closer together in time and space are more highly correlated than data further apart.

b. Homoscedasticity: The variance of the errors should not be a function of any of the predictor variables. If the residuals are heteroscedastic, occasionally the dependent variable can be transformed to preserve homoscedasticity.

To test hypotheses concerning the regression equation, the assumption must be added that the errors are normally distributed. These assumptions of independence, homoscedasticity, and normality can be empirically tested (Bock, 1975). While violations of the assumptions of homoscedasticity and normality do not seriously distort the level of significance, violation of the independence assumption does.

To test whether a given beta weight is not zero, compute the multiple correlation without the variable (R_1) and with the variable (R_2) in the equation. Then under the null hypothesis that the coefficient is zero the following is distributed as

$$t(N-k-1) = \left[\frac{(R_2^2 - R_1^2)(N - k - 1)}{1 - R_2^2}\right]^{1/2} \quad [4.5]$$

where N is the total number of sampling units, k is the number of predictors in the full equation, and t is Student's t distribution. The number of degrees of freedom is $N - k - 1$. To test whether there is an effect of father's occupation, first omit it from the equation and R_1 is simply the correlation of I with G, or .572. Since $N = 767$, there are 764 degrees of freedom and earlier R_2 was found to be .582. Substituting these values into Equation 4.5 one obtains

$$t(764) = \left[\frac{(.582^2 - .572^2)764}{1 - .582^2}\right]^{1/2} = 3.65$$

which is statistically significant at the .05 level. Similarly, the test that the effect of intelligence is nil is

$$t(764) = \left[\frac{(.582^2 - .248^2)764}{1 - .582^2}\right]^{1/2} = 17.90$$

which is also significant.

If two or more predictors are removed from the regression equation, a similar procedure is followed. Compute R_1 omitting the variables that one wishes to exclude and R_2 for the full equation. It can then be shown that under the null hypothesis of zero coefficients, the following is distributed as

$$F(m,N-k-1) = \left[\frac{R_2^2 - R_1^2}{1 - R_2^2}\right]\left[\frac{N - k - 1}{m}\right] \quad [4.6]$$

where m equals the number of excluded variables and F is Fisher's F distribution. The F of Equation 4.6 has m degrees of freedom in the numerator and $N - k - 1$ in the denominator. To test the effect of removing both father's occupation and intelligence, the R_1 value is zero since there are no predictor variables. Substituting into Equation 4.6 one obtains

$$F(2,764) = \frac{(.582^2 - .000^2)764}{(1 - .582^2)2} = 195.67$$

which is highly significant. Testing whether all the beta weights are zero is identical to testing whether the multiple correlation is zero.

A beta weight with only one predictor is identical to a correlation coefficient. A beta weight with two or more predictors is not a correlation and can even be larger than one. For instance,

$$X_3 = .75X_1 + 1.25X_2$$

where $r_{12} = -.6$ is a perfectly valid regression equation. (As an exercise show $r_{13} = 0$ and $r_{23} = .8$.) Although beta weights larger than one are not impossible, they are empirically improbable. One should be suspicious when one obtains weights larger than one.

Beta weights or path coefficients can be easily restated in an unstandardized metric. For instance, given

$$X_3' = aX_1' + bX_2' + cU'$$

where the prime notation expresses standardization, one can reexpress a and b in an unstandardized metric: The unstandardized coefficient of X_1 is $a[V(X_3)/V(X_1)]^{1/2}$, and of X_2, $b[V(X_3)/V(X_2)]^{1/2}$. Thus, to destandardize a coefficient, multiply it by the standard deviation of the endogenous variable and divide it by the standard deviation of the exogenous variable. To standardize an unstandardized coefficient, reverse the operation: Multiply it by the standard deviation of the exogenous variable and divide it by the standard deviation of the endogenous variable.

THE CAUSAL INTERPRETATION OF REGRESSION COEFFICIENTS

Researchers are often interested not only in predicting grades but also in explaining what causes grades. Is it permissible to interpret the regression coefficients obtained from the regression analysis as causal parameters? The answer is, as it is in the case of most important questions, it depends. Regression coefficients can be interpreted as causal coefficients if certain assumptions are met. These assumptions are the same as those of multiple regression: In this case the justification is not to maximize prediction nor to perform significance tests, but because the assumptions are specified by theory. The assumption of independence can usually be assured by the sampling design of the research. Homoscedasticity and normality of errors are usually not discussed in most social science theories (although they are in some biological theories), but as stated earlier, these are robust assumptions which may only be "approximately" met. Often a transformation of the measure into a more meaningful metric will aid in meeting these two assumptions.

As a result of least squares, the residuals are uncorrelated with the predictor variables. In structural modeling, this result requires that the disturbance be uncorrelated with the exogenous variables. As shall be seen, the assumption of *uncorrelated errors* implies that:

a. The endogenous variable must not cause any exogenous variable; that is, there is no reverse causation.

b. The exogenous variables must be measured without error and with perfect validity.

c. None of the unmeasured causes must cause any of the exogenous variables; that is, there are no common causes, or *third variables*.

Obviously, lack of reverse causation, perfect measurement, and lack of common causes are rather stringent assumptions. Reverse causation can be ruled out by theory or logic; for example, variables measured at one point in time do not cause variables measured earlier in time. If reverse causation cannot be ruled out, however, a nonhierarchical model must be specified the parameters of which cannot be estimated by an ordinary regression analysis (see Chapter 6). Perfect measurement is assured for some variables such as race and sex, close approximations exist for some variables such as intelligence, but most motivational, attitudinal, and behavioral measures are not close to being measured perfectly. The common cause problem can be solved by measuring the third variables (although these variables must be perfectly measured), but it is still logically impossible to demonstrate that all third variables have been excluded if the multiple correlation is less than one. The next chapter presents partial solutions to the problems of unreliability and unmeasured third variables.

A final assumption for the application of regression analysis to structural models *when the variables are measured cross-sectionally* is that a state of equilibrium has been reached. Thus if X_1 is assumed to cause X_2 with a lag of k units, and if X_1 and X_2 are contemporaneously measured at time t, equilibrium exists if $X_{1t} = X_{1t-k}$; that is, X_1 did not change between times $t - k$ and t.

The path coefficients are, then, estimated by the beta weights, and the causal coefficient for the disturbance of the exogenous variable simply equals $(1 - R^2)^{1/2}$, where R^2 is the squared multiple correlation of the regression equation. In some treatments of path analysis, the disturbance has no path coefficient. For this case the variance of the disturbance is set at $1 - R^2$. Here the variance of the disturbance is fixed at one and its path coefficient is estimated, thereby making all of the variables of the model standardized. Actually, $(1 - R^2)^{1/2}$ is a biased estimate of the disturbance path since R^2 is also biased. A less biased estimate of the multiple correlation is $R^2(N - k - 1)/(N - 1)$. One could then use this corrected R^2 to estimate the disturbance path. For ease of computation here the uncorrected estimate is used. However, if sample size is small, the corrected R^2 should be used.

To illustrate the use of multiple regression to estimate causal parameters, return to the three variables. A simple model is

$$I = aF + bU \quad [4.7]$$

$$G = cF + dI + eV \quad [4.8]$$

That is, father's occupation (F) causes intelligence (I), and father's occupation and intelligence both cause school grades (G). It is assumed $\rho_{UF} = \rho_{VF} = \rho_{VI} = 0$; that is, the disturbances are uncorrelated with the exogenous variables, which in turn implies $\rho_{UV} = 0$. (As an exercise, prove why $\rho_{UV} = 0$.) The path diagram for Equations 4.7 and 4.8 is in Figure 4.1.

The path from F to I can be estimated by the regression of I on F since the disturbance of I is uncorrelated with F. The univariate beta weight is simply the correlation coefficient. To see this, solve for r_{FI} by either the first law or the tracing rule, to obtain a, the path from I to F. Parameter a then is estimated by .250. To test whether such a value could be explained by sampling error, use Equation 4.5

$$t(765) = \left[\frac{(.250)^2 765}{.9375}\right]^{1/2} = 7.14$$

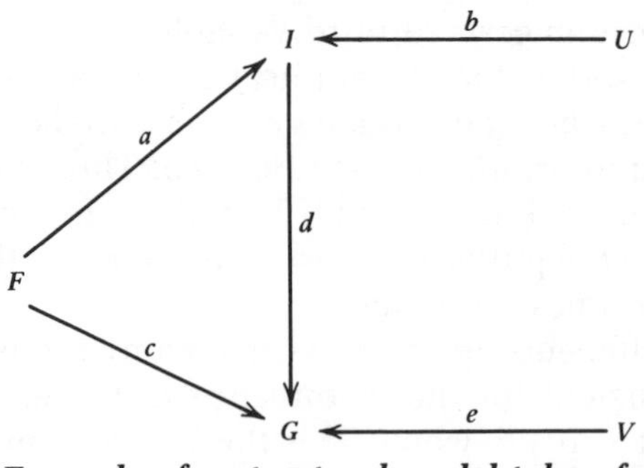

I b U
a d
F
c
G e V

Figure 4.1 Example of a structural model taken from Kerchoff.

(Note $R_1 = 0$ since no variables remain in the equation.) The path for the disturbance equals $(1 - R^2)^{1/2}$ where R^2 is the multiple correlation. Since there is only one predictor, and $R = .250$, $b = (1 - .250^2)^{1/2} = .968$. To solve for c and d, regress G on F and I which was done earlier in this chapter. Thus, the estimate of c equals .112 and of d equals .544. Also computed was the multiple correlation for that equation and obtained was the value of .582. The estimate of e is then $(1 - .582^2)^{1/2} = .813$. Earlier these beta weights were shown to be statistically significant.

It is just that simple. If the disturbances are uncorrelated with exogenous variables, then beta weights estimate path coefficients and $(1 - R^2)^{1/2}$ estimates the path from the disturbance to the endogenous variable. Tests of hypotheses about the path coefficients can be obtained from ordinary least squares (standard multiple regression analysis).

Note that in applying regression analysis to estimate structural parameters, one estimates a single regression equation. The use of stepwise regression has no place in structural modeling. With stepwise regression one ordinarily does not know which variables are important, but in structural modeling, theory has already specified the important variables.

Causal Linkages

For the model in Figure 4.1, F causes G both directly and indirectly. This can be seen by examining r_{GF} using the tracing rule:

$$r_{GF} = c + ad$$

There is a direct linkage, path c, and an indirect linkage, F causes I, and, in turn, I causes G. The variable I *mediates* a link or intervenes between F and G. The mathematics of indirect paths are discussed later in this chapter.

The distinction between direct and indirect causes is an important one in social science. When a causal link has been demonstrated either experimentally or nonexperimentally, usually some inquiring person will investigate the variables that *mediate* that causal linkage; that is, an attempt will be made to show that X causes Y through Z and that once Z has been controlled, the relationship between X and Y vanishes. There are usually a host of mediating variables, and through such a process variables in social science become refined and theory expands. One could argue that all causal linkages are indirect, or mediated, thereby making a true direct link an impossibility. A causal linkage, therefore, is said to be direct only with respect to other variables specified in the system under consideration. Certain classes of variables are not entered into structural equations generally because they are not topics of study in that field. For instance, it is likely that physiological variables mediate the psychological, and physical laws mediate certain chemical laws. Researchers do not strive to find absolute linkages, but rather look for linkages relative to their interests and discipline.

Returning to Figure 4.1, I is related to G in two ways: by a direct causal linkage and by a *spurious* linkage through F since F causes I and G. A spurious relationship means that the association between two variables is a result of both being caused by a third variable. Although all indirect causes need not be entered into the structural equation, spurious variables must be. For instance, since the linkage of I and G through F in Figure 4.1 does not reflect the causal effect of I on G, the variable F must be measured and controlled. Care must be taken to specify from theory these spurious causes.

Although it makes no substantive sense in this example to say that I causes F, even if it were so, the path coefficients of the effect of I and F on G (c and d) would not change. Similarly, if no model for the relationship between the exogenous variables is posited and a curved line is drawn between them, the estimates of c and d again remain the same. This result follows from the earlier rule that multiple regression correctly estimates structural coefficients if the disturbances are uncorrelated with the specified causes. Nothing need be said about the causal relationships among the exogenous variables. Even if the structural model among the exogenous variables contains specification error, estimates of the paths to the endogenous variables will not be affected by this error.

Partial Correlation

Consider a different causal model for our three variables:

$$I = aF + bU$$

$$G = cF + dV$$

where $e = \rho_{UV} \neq 0$ and $\rho_{UF} = \rho_{VF} = 0$. As with the model in Figure 4.1, this model assumes F causes I and G. But in this case the disturbances U and V are correlated because various common causes have been omitted. Since there are three free parameters, a, c, and e, and three correlations, the parameters may be just-identified. From the path dia-

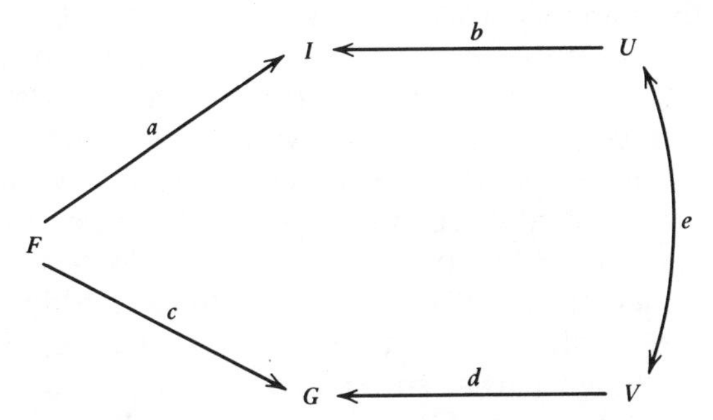

Figure 4.2 Correlated disturbances.

gram in Figure 4.2, the three correlations obtained by the tracing rule are:

$$r_{FI} = a$$

$$r_{FG} = c$$

$$r_{IG} = ac + bde \qquad [4.9]$$

Path coefficients a and c are directly estimated by the correlations. Parameters b and d can be solved from the multiple correlations

$$b = (1 - r_{FI}^2)^{1/2} = .968$$

$$d = (1 - r_{FG}^2)^{1/2} = .969$$

Since a, b, c, and d are known, one can solve for e from Equation 4.9:

$$e = \frac{r_{IG} - ac}{bd}$$

$$= \frac{r_{IG} - r_{IF}r_{GF}}{((1 - r_{IF}^2)(1 - r_{GF}^2))^{1/2}} \qquad [4.10]$$

Solving through obtains .544 as an estimate of e. Some readers may recognize Equation 4.10 as the *partial correlation* of I and G with F partialled out, or as it is symbolized $r_{IG \cdot F}$. A partial correlation can be tested against the null hypothesis that it equals zero by the following formula:

$$t(N-k-2) = \frac{r_p(N - k - 2)^{1/2}}{(1 - r_p^2)^{1/2}} \qquad [4.11]$$

where r_p is the partial correlation and k is the number of variables partialled out. (The preceding formula is no different from the standard test of a correlation.) Testing the estimate of e yields a $t(764) = 17.92$. If one goes back to compare the test of significance with the test of d in Figure 4.1, one notes that both yield identical values within rounding error. This is no accident since they are equivalent. Blalock (1964) had originally suggested using partial correlations to test assumptions about path coefficients. A better approach is to use multiple regression which besides providing significance tests also estimates the structural coefficients. Although partial correlations and regression coefficients may be related, the choice of which to compute depends on the substantive meaning the researcher wishes to attach to a coefficient. A partial correlation is the correlation between the unspecified causes of two endogenous variables, while a regression coefficient is the path from an exogenous to an endogenous variable.

At present, for the correlation between disturbances to be identified, it must be assumed that neither endogenous variable causes the other. For instance, a path drawn from I to G in Figure 4.2 would create an underidentified model since there are four free parameters and only three correlations. Thus, to estimate the correlation between disturbances, the assumption must be made that neither endogenous variable causes the other, that is, neither variable appears in the equation of the other.

Many texts give the formula for higher order partials and computer programs will compute them. If you are caught without a computer, remember that the partial is the correlation between residuals from two regression equations using the same predictor variables. One need only compute the covariance between residuals and their variances. The actual residuals need not be computed, but only their theoretical variances and covariance must be.

Semipartial Correlation

There is yet another model for the three variables. For this model the disturbance is a cause of two endogenous variables:

$$I = aF + bU$$

$$G = cF + dU + eV$$

where $\rho_{UF} = \rho_{FV} = \rho_{UV} = 0$. The correlations can be obtained by applying the tracing rule to the path diagram in Figure 4.3. They are

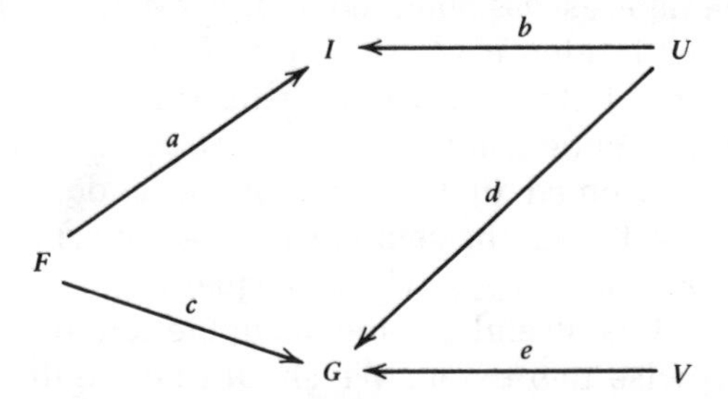

Figure 4.3 Path from a disturbance to two endogenous variables.

$$r_{IF} = a$$

$$r_{GF} = c$$

$$r_{GI} = ac + bd$$

The solution for both a and c is directly given again. The estimate b is $(1 - r_{IF}^2)^{1/2}$, or .968. From the correlation between G and I one can solve for d:

$$d = \frac{r_{GI} - ac}{b}$$

$$= \frac{r_{GI} - r_{IF}r_{GF}}{(1 - r_{IF}^2)^{1/2}}$$

which is the *semipartial correlation*, or *part correlation*, between G and I controlling for F, or as it is symbolized $r_{G(I \cdot F)}$. A test of the semipartial is given by

$$t(N-m-1) = \frac{r_s(N - m - 1)^{1/2}}{(1 - R^2)^{1/2}} \qquad [4.12]$$

where r_s is the semipartial, m is the number of causes of the variable that is caused by the disturbance (e.g., G in Figure 4.3), and R^2 is the multiple correlation of that variable. This R^2 can be simply computed by first computing R^2 with the causal disturbance omitted and then adding to it the squared semipartial. For instance, for Figure 4.3 d is estimated by the value of .527, R^2 is then $.248^2 + .527^2 = .339$, and the test of significance is

$$t(764) = \frac{.527(764)^{1/2}}{(1 - .339)^{1/2}} = 17.92$$

Again this is the same as was obtained for the test of significance of d in Figure 4.1 and of e in Figure 4.2. However, the interpretation is different. The semipartial is the effect of a disturbance of one endogenous variable on another endogenous variable. At present, to identify this path it must be assumed that neither of the endogenous variables causes the other. Note that the semipartial is a nonsymmetric measure of association. Ordinarily $r_{1(2 \cdot 3)}$ will not equal $r_{2(1 \cdot 3)}$.

The semipartial is useful in the computation of regression coefficients in stepwise regression. We should not confuse this usefulness with its causal interpretation. The semipartial has not and probably should not have much role in structural analysis.

Example

It has been shown that beta coefficients estimate path coefficients if the unspecified causes are uncorrelated with the exogenous variables, that partial correlations estimate the correlation of the disturbances of two endogenous variables if neither variable appears in the equation of the other, and that the semipartial estimates a path of a disturbance to another endogenous variable, again if neither endogenous variable appears in the equation of the other. Here these principles are applied to determine the statistical methods for estimating causal parameters of the four models in Figure 4.4. All four models have in common the characteristic that they are just-identified. Each model has four observed variables, and therefore six correlations; and since each model has six free parameters, each model may be just-identified. Those six parameters can be identified as follows:

Model I: Parameters a, b, and c are simply estimated by the correlations between the purely exogenous variables. To estimate d, e, and f regress X_4 on X_1, X_2, and X_3, and the resulting beta weights are estimates of path coefficients. Parameter g is $(1 - R^2)^{1/2}$, where R is the multiple correlation of X_4 with $dX_1 + eX_2 + fX_3$.

Model II: Parameter a is estimated by regressing X_2 on X_1. Parameters b and c are estimated by regressing X_3 on X_1 and X_2, and d, e, and f by regressing X_4 on X_1, X_2, and X_3. The estimates of d, e, and f are the same for Models I and II. The difference between the two models is that Model I makes no statement about the causal relationships between X_1, X_2, and X_3, while Model II does. The disturbances of g, h, and i can be estimated by $(1 - R^2)^{1/2}$, where R is the multiple correlation for each equation.

Model III: To estimate a regress X_2 on X_1; b, regress X_3 on X_1; and c, regress X_4 on X_1. Since there is only one predictor variable in each

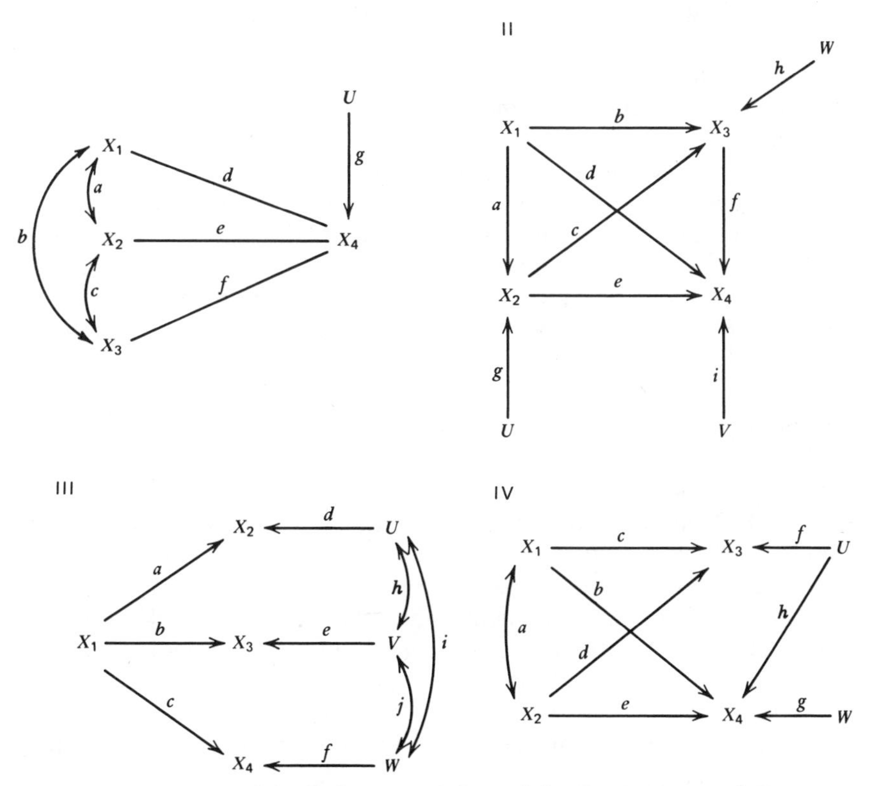

Figure 4.4 Just-identified regression models (X_1, X_2, X_3, and X_4 are measured).

equation, the estimated causal coefficient equals the correlation of cause and effect. Parameters h, i, and j can be estimated by partial correlations since X_2, X_3, and X_4 do not cause each other: $h = r_{23\cdot1}$, $i = r_{24\cdot1}$, and $j = r_{34\cdot1}$. The disturbances can be estimated by $(1 - R^2)^{1/2}$.

Model IV: Parameter a is simply estimated by r_{12}. Parameters c and d can be estimated by regressing X_3 on X_1 and X_2, and b and e by regressing X_4 on X_1 and X_2. Since X_3 does not cause X_4 or vice versa, h is simply the semipartial correlation of X_3 with X_4 controlling for X_1 and X_2—$r_{3(4\cdot12)}$. The paths for residuals are estimated in the usual way. Remember to include U as a cause of X_4 in the multiple correlation for X_4.

The beta weights of each of the preceding four models can be tested against the null hypothesis that they are zero by applying the t test given by Equation 4.5, the partial correlations in Model III by the t test given by Equation 4.11, and the semipartial in Model IV by Equation 4.12.

After estimating the coefficients and testing them against zero, one begins what is perhaps the most interesting part of causal analysis. If certain beta coefficients are not significantly different from zero, the researcher should question the utility of a path there. But as in any significance test there may be errors of inference. There may actually be a path but either the sample size is too small or the exogenous variable is too highly correlated with the other exogenous variables. (This second problem is called *multicollinearity*, a problem examined later in this chapter.) But given high power, a zero path should make the researcher question the validity of a causal linkage, at least for the population at hand.

The researcher should also examine the size (high, medium, or low) and the direction (positive or negative) of the paths. If they are anomalous, then the researcher should question the specification of the model. For instance, if one were to find that the path from actual similarity to attraction is negative, one should suspect that something is wrong with the model. Unfortunately, it may be necessary to estimate many different causal models and find the model that makes the best sense. Obviously, inference is stronger if only one model is stated a priori (without looking at the data) and data confirm that model than if a model is fitted to the data. Causal modeling is meant to be a *confirmatory* method (Tukey, 1969) but we often find ourselves using it in an exploratory mode.

Even when causal modeling is perfectly applied, it can only confirm or disconfirm that a given model fits the correlations; it never proves the model. An infinite number of models can fit any set of correlations. Ideally the researcher designs the study in such a way that two or more theoretically opposing causal models have divergent implications for the correlations. To increase the validity of a causal model, it is useful to add to the structural equations a variable that one theory says is a cause and another theory says is not. If the first theory is correct, then the relevant path should be nonzero, while for the other theory it would be zero. Unfortunately, researchers are reluctant to spend their time measuring variables that they do not believe to be important. But a model can be strengthened if certain variables show the desired zero paths and other variables the desired nonzero paths. This gives the model both *convergent* and *discriminant* validity (see Chapter 13); that is, theory is confirmed in terms of both what should and what should not happen.

To illustrate this point, assume that a researcher did not believe that there was a path from X_2 to X_3 for Model II in Figure 4.4. The researcher should still estimate the path, and test whether it is zero. Along the same lines, in Model IV, if the researcher did not believe that there was a path from U to X_4, the researcher should still estimate the path by $r_{3(4 \cdot 12)}$ and test whether it significantly differs from zero.

TECHNICAL ISSUES

Overidentification

Returning to Figure 4.4, for Model II, if the path labeled c is deleted, the model is overidentified. There is one less parameter than correlation. The early path analysts originally suggested a seemingly sensible strategy for overidentification. Examine the model in Figure 4.5. The model is overidentified since there are three correlations and only two free parameters. As with many overidentified models, it is possible to obtain two estimates of all the parameters of the model: $a = r_{12}$ and $a = r_{13}/r_{23}$, and $b = r_{23}$ and $b = r_{13}/r_{12}$. Setting both the estimates equal and restating in terms of population parameters yields the overidentifying restriction of

$$\rho_{13} - \rho_{12}\rho_{23} = 0 \qquad [4.13]$$

Since there are two estimates of a parameter, how should the two estimates be pooled? A simple solution would be to add the two together and divide by two. Goldberger (1970) has shown that the statistically best solution is to use the least-squares solution and ignore any other solution. The least-squares solution is more efficient than a pooled solution. Thus, if the theory is correct, one should estimate the path coefficients by regression analysis yielding r_{12} as an estimate of a and r_{23} as an estimate of b. I would not argue for exactly this solution. It is possible that there is a specification error of a path from X_1 to X_3. If

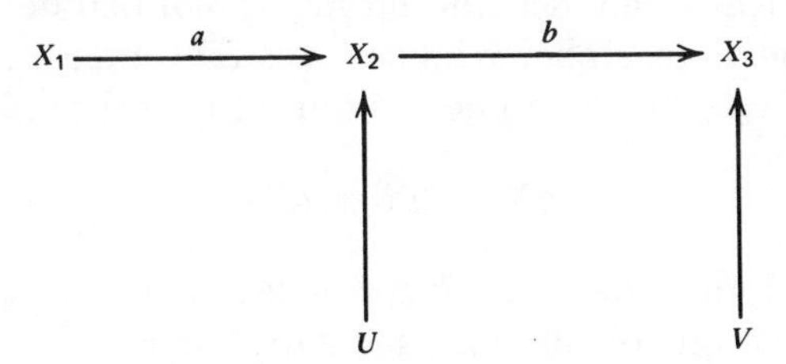

Figure 4.5 Causal chain.

there is such a path, the estimate of the path from X_2 to X_3 will be distorted through this specification error. It might be wiser to regress X_3 on both X_1 and X_2 and to expect the path from X_1 to X_3 to be nonsignificant. The estimate of that path is

$$\frac{r_{13} - r_{12}r_{23}}{1 - r_{12}^2}$$

Note that for the preceding to equal zero, Equation 4.13 must hold. Thus, the overidentifying restriction is tested by the prediction of the zero path. A price is paid for this strategy. The variance of the estimate of b is increased, but at least the researcher knows that b is not biased if X_1 causes X_3. We pay for less bias by decreased efficiency.

It is my opinion that for models for which regression is appropriate, the model should always be just-identified; that is, as many parameters as correlations should be estimated. Theory will specify which parameters will be zero and nonzero. The researcher also has a choice of what connection to have among the endogenous variables. For instance, for Model IV in Figure 4.4, if there were no path from U to X_4 specified, there would be a number of alternatives for adding another parameter. One could have (a) a path from X_3 to X_4, (b) a path from X_4 to X_3, (c) a path from U to X_4, (d) a path from V to X_3, or (e) correlated disturbances between U and V. Whatever additional parameter one chooses to estimate, the null hypothesis of a zero value tests the same overidentifying restriction and the same t statistic is obtained.

Specification Error

Too often researchers examine the simple, or raw, correlation coefficient as an indication of causal effects. The naive logic is that if X causes Y, then X and Y should be correlated, and if X does not cause Y, they should be uncorrelated. Neither statement is true. After controlling for other exogenous variables, a strong relationship can vanish and a zero relationship can become strong. To illustrate both of these effects, imagine a researcher who believes that three variables, X_1, X_2, and X_3, cause X_4. The researcher's implicit causal model is

$$X_4 = aX_1 + bX_2 + cX_3 + dU$$

with the assumption that U is uncorrelated with X_1, X_2, and X_3. Suppose the correlations of three causes with X_4 are

$$r_{14} = .00$$

$$r_{24} = .45$$

$$r_{34} = .45$$

Given these correlations, the researcher immediately jumps to the conclusion that X_2 and X_3 are relatively strong causes of X_4 while X_1 has no effect. We, however, would know better. A correlation between a hypothetical cause and effect is a function of not only the causal effect, but also the correlation of the cause with the other causes. This can be seen by applying the first law:

$$r_{14} = a + br_{12} + cr_{13}$$

$$r_{24} = ar_{12} + b + cr_{23}$$

$$r_{34} = ar_{13} + br_{23} + c$$

Knowing that $r_{12} = -.3$, $r_{13} = .5$, and $r_{23} = .6$ allows one to solve for a, b, and c by multiple regression. The resultant values are $a = -.3$, $b = .0$, and $c = .6$. So, although X_1 is uncorrelated with X_4, X_1 still has an effect on X_4! (X_1 is sometimes called a supressor variable [Lord & Novick, 1968, pp. 271–272].) Also, although X_2 is correlated with X_4, its causal effect is zero! Clearly one must go beyond the simple raw correlation to make causal interpretations.

What happens to the estimates of causal parameters if the researcher omits one of the exogenous variables? If either X_1 or X_3 are omitted, the other exogenous variables' path coefficients are biased since the variables are correlated with disturbances, and thus the model contains specification error. In Table 4.2 are the estimates of path coefficients with one exogenous variable omitted. If X_2 is omitted, the estimates of the path coefficients are unchanged since X_1 and X_3 remain uncorrelated with the disturbances. As was stated earlier, including variables

Table 4.2. Path Coefficients with One Omitted Variable

	Omitted Variable			
Path Coefficient	X_1	X_2	X_3	**None**
a	—	−.300	.148	−.300
b	.281	—	.495	.000
c	.281	.600	—	.600

that have no causal effects in no way biases the estimates of path coefficients. In fact, it adds increased validity to the model by making an explicit prediction of a zero path. Returning to Table 4.2, if the model contains specification error by omitting either X_1 or X_3, the causal coefficients become misleading. For example, when X_3 is omitted, the path from X_1 to X_4 is .148, when it should be $-.3$. This bias is due to X_1 being correlated with the disturbance since X_3, which causes X_4, is contained in the disturbance and is correlated with X_1.

To see more formally the effect of an omitted variable, examine the path diagram in Figure 4.6. If variable X_3 is omitted then the estimate of d is

$$\frac{\rho_{14} - \rho_{12}\rho_{24}}{1 - \rho_{12}^2} = \frac{d + \rho_{12}e + \rho_{13}f - \rho_{12}(e + \rho_{12}d + \rho_{23}f)}{1 - \rho_{12}^2}$$

$$= \frac{d(1 - \rho_{12}^2) + f(\rho_{13} - \rho_{12}\rho_{23})}{1 - \rho_{12}^2}$$

$$= d + f\beta_{31\cdot 2}$$

Thus there is no bias in estimating the effect of an exogenous variable if either the omitted variable does not cause the endogenous variable or the regression of the omitted variable on the exogenous variable controlling for the remaining exogenous variables is zero.

An example of the necessity of controlling for variables can be taken from Fine (1976). He compared the ability of psychics and nonpsychics to predict the future. He took 59 predictions of psychics from the *National Enquirer* and 61 predictions from nonpsychics. He then had observers rate how likely the predictions were to occur. A year later he had a different set of observers judge whether the events had occurred

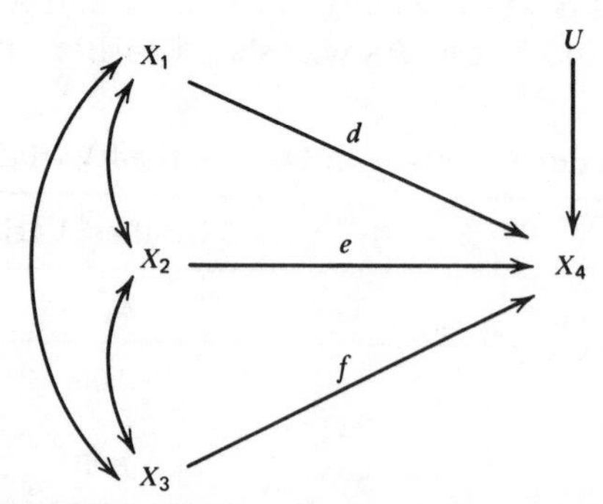

Figure 4.6 The omitted variable problem.

during the previous year. There are then three variables: psychic–nonpsychic (P), likelihood of outcome (L), and actual outcome (A). The psychic-nonpsychic variable is a dummy variable in which one is assigned for psychic and zero for nonpsychic. The sample size is 120, the total number of predictions.

The correlations among the three variables are

$$r_{PL} = -.440$$

$$r_{PA} = -.329$$

$$r_{LA} = .572$$

The simple correlation of being a psychic with actual outcome is highly significant, $t(118) = -3.78$, indicating psychics did poorer than nonpsychics. Estimating the following structural equation:

$$A = aP + bL + cU$$

yields an estimate a to be $-.095$, b to be .530, and c to be .816. Thus, once one has controlled for the likelihood of the outcome, the difference between psychics and nonpsychics is reduced substantially. The test of a is

$$t(117) = \left[\frac{(.335 - .327)(117)}{(.665)(1)}\right]^{1/2}$$

$$= 1.14$$

and of b

$$t(117) = \left[\frac{(.335 - .108)(117)}{(.665)(1)}\right]^{1/2}$$

$$= 6.31$$

and thus the effect of being a psychic is not significant. Even though being a psychic correlates negatively with the outcome, it correlates about as negatively as it should, given that psychics make predictions that are unlikely and unlikely predictions do not usually come true.

Multicollinearity

Multiple regression estimates the effect of a predictor variable on a criterion taking into account both the correlations between the predictor variable and the other predictor variables and the *effects* of the other predictor variables. As is explained in Chapter 10, for classical experimental design the predictors, or independent variables, are uncorrelated when equal numbers of subjects are assigned to each cell. If the independent variables are uncorrelated, a simpler, more elegant computational strategy can be employed: the analysis of variance. However, the strategy of partitioning of variance makes little or no sense in the case when the independent variables are correlated (Duncan, 1970, 1975). As the exogenous variables become more and more intercorrelated, the standard errors of path coefficients increase. This increased uncertainty in regression coefficients due to intercorrelated predictor variables is called *multicollinearity*. Increased standard errors means lower power. Gordon (1968) provides an excellent review of some of the obvious and not so obvious effects of multicollinearity. Much has been made of the "problem of multicollinearity," but I prefer to see it not so much a problem as a *cost* that is paid by not employing an experimental factorial design with equal cell size (see Chapter 10). It may be a cost that is well worth paying to increase external validity (Campbell & Stanley, 1963).

Before discussing the effects of multicollinearity, its measurement should be first reviewed. Most measures of multicollinearity center around the multiple correlation of a predictor variable with the other predictors. A helpful fact not usually mentioned in most multivariate texts is that the diagonal elements of the inverse of the correlation matrix are $1/(1 - R_i^2)$, where R_i is the multiple correlation of variable i with the other predictor variables. Some computer programs output *tolerance* which is usually defined as $1 - R_i^2$, where R_i is defined as previously (Harris, 1975, p. 283). As R_i^2 increases, multicollinearity becomes more of an issue.

Multicollinearity is literally built into a set of structural equations. If X_1 causes X_2 and X_2 causes X_3, it is all but inevitable that X_1 and X_2 are correlated. If X_1 is a strong cause of X_2, it may be difficult, if not impossible, to disentangle the causal effects of X_1 and X_2 on X_3 with a small sample. For example, in Figure 4.5, if a equaled .87 and b equaled .20, one would need close to 783 units to have a 80% chance of finding the estimate of a statistically significant at the .05 level. However, if a were zero and b were .20, one would need only 193 units to have 80% power (cf. Cohen, 1969, p. 99). Thus, causal effects among the exogenous variables tend to reduce power through multicollinearity.

Multicollinearity also enters structural analysis when two or more measures of the same cause are employed. Imagine that

$$X_2 = .3X_1 + .954U$$

where $\rho_{1U} = 0$. Suppose the researcher does not measure X_1 but has two errorful measures of X_1:

$$X_3 = .9X_1 + .436E$$

$$X_4 = .9X_1 + .436F$$

where $\rho_{1E} = \rho_{1F} = \rho_{EU} = \rho_{FU} = 0$, and $\rho_{EF} = 0$. The path model for the three equations is in Figure 4.7. The correlations between the measured variables are $\rho_{34} = .81$, $\rho_{23} = .27$, and $\rho_{24} = .27$. Now suppose the researcher estimates the following equation:

$$X_2 = aX_3 + bX_4 + cV$$

The estimates for a and b are both .149 and the multiple correlation is is .284 which approaches .3. The surprising fact is that whereas the theoretical .27 correlation of either X_3 and X_4 with X_2 needs only 54 units to be significant about 50% of the time, the theoretical beta weight of .149 needs 474 units to be significant about 50% of the time, again using the .05 level of significance. This is a case in which two is not better than one. Adding another measure of a cause increases the multiple correlation but it tends both to make beta weights smaller and to make the standard errors of the beta weights larger.

As this example illustrates, special care should be taken not to include in the regression equation more than one measure of the same construct. I do not mean to imply that a construct should not have more than one measure. Quite the opposite, the multiple indicator strategy is often useful, as is demonstrated in Chapter 8. But the researcher should

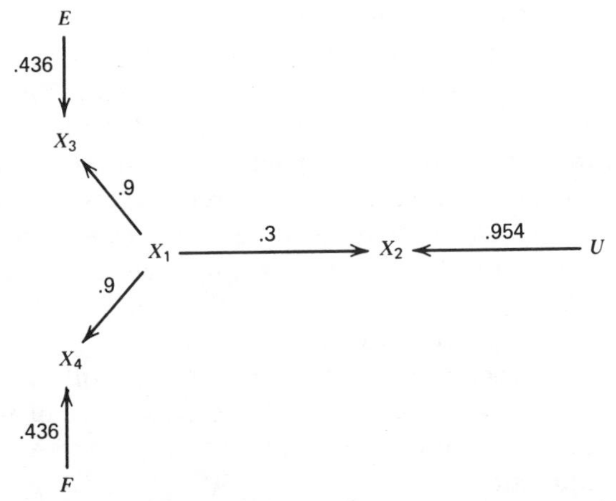

Figure 4.7 Two indicators of an exogenous variable.

be suspicious of highly correlated exogenous variables in a regression equation, since it may well be that they are multiple indicators of the same cause. It is not sufficient for the researcher simply to define two variables as measures of different constructs. If the two variables correlate as highly as the average of their reliabilities, then the measures have no discriminant validity (Campbell & Fiske, 1959). It is not enough to say that measures tap different constructs; it should be demonstrated empirically.

Multicollinearity becomes an insuperable problem when the correlation between two exogenous variables is unity. To see this, examine the following structural equation:

$$X_3 = aX_1 + bX_2 + cU$$

where $r_{1U} = r_{2U} = 0$. By the first law the correlations are

$$r_{13} = a + br_{12}$$

$$r_{23} = ar_{12} + b$$

Ordinarily the researcher would solve for a and b, but if $r_{12} = 1$, then $r_{13} = r_{23} = a + b$. There is a single equation left in two unknowns for which there is no solution. This is an example of empirical underidentification.

Much to our despair and happiness, we rarely obtain correlations or multiple correlations of one; but we can bring about perfect multicollinearity by not being careful. First, as the number of predictors approaches the number of units, multicollinearity increases. Having more predictors than subjects insures perfect multicollinearity. Second, careless creation of linear composites can result in perfect multicollinearity. Say, for instance, a researcher is interested in the effects of power in heterosexual relationships. The researcher obtains a measure of the resources of the male and the resources of the female. Power in the relationship is then operationalized as the difference between the male's and female's resources. The researcher regresses the involvement of one partner on the two resource variables and power. To the researcher's surprise the computer program blows up and no estimates are possible. This is due to a linear dependency in the regression analysis. Knowing any two of the measures perfectly predicts the third. Probably the best procedure for the researcher to follow is to regress involvement on the two resource variables. The test of the effect of power is that the regression coefficients for the resource measures are significantly different.

Equality of Regression Coefficients

In the previous section it was mentioned that to test the effect of power in heterosexual relationships it should be tested whether the coefficient for male's resources equaled the coefficient for female's resources. It has not, however, been shown how to test the equality of two coefficients within the same equation. The test is relatively simple. Imagine for Equation 4.8 that it is assumed that c equals d. This would imply

$$G = c(F + I) + eV$$

Thus, to test whether two coefficients are equal compare the R^2 of an equation with both exogenous variables in the equation with the R^2 of the sum of the two in the equation. If the parameters are equal, nothing should be gained by estimating separate parameters.

For the particular example

$$\begin{aligned} V(F + I) &= V(F) + V(I) + 2C(F,I) \\ &= 1 + 1 + 2(.25) \\ &= 2.5 \end{aligned}$$

and

$$\begin{aligned} C(F + I,G) &= C(F,G) + C(I,G) \\ &= .572 + .248 \\ &= .820 \end{aligned}$$

This makes $r_{F+I,G} = .82/(2.5)^{1/2} = .519$. Since $(F + I)$ is the only predictor of G, the path from $(F + I)$ to G is also .519 as is the multiple correlation. Recall that the multiple correlation with both F and I in the equation is .582. The test of whether computing both c and d separately significantly increases R^2 by Equation 4.5 is

$$\begin{aligned} t(764) &= \left[\frac{(.582^2 - .519^2)764}{(1 - .582^2)}\right]^{1/2} \\ &= 8.95 \end{aligned}$$

which is highly significant. A researcher should carefully consider whether the assumption of the equality of regression coefficients refers to the standardized or unstandardized coefficient. *Ordinarily it refers to the unstandardized coefficient.*

Tests of the equality of the same coefficients in the same equation computed from two samples is given in Snedecor and Cochran (1967). More complicated tests of the equality of coefficients is given in Werts, Rock, Linn, and Jöreskog (1976).

Measuring Indirect Effects

Earlier it was stated that an exogenous variable can have a direct and an indirect effect. For instance, for the path model in Figure 4.8 of the Kerchoff data in Table 4.1, intelligence is a direct and an indirect cause of occupational aspiration. *The total effect equals the direct effect plus the sum of the indirect effects.* The direct effect is given by the path coefficient. Some early path analysts stated that the total effect is simply the correlation between the endogenous and the exogenous variable. However, a little reflection shows that the correlation between variables reflects more than direct and indirect effects.

If the causal variable is purely exogenous, then part of its correlation with the endogenous variable may be due to its correlation with the other purely exogenous variables and the effects of these purely exogenous variables. To see this let us examine the correlation of intelligence (I) with grades (G):

$$r_{IG} = p_{GI} + p_{GS}r_{IS} + p_{GE}r_{IE} + p_{GF}r_{IF}$$

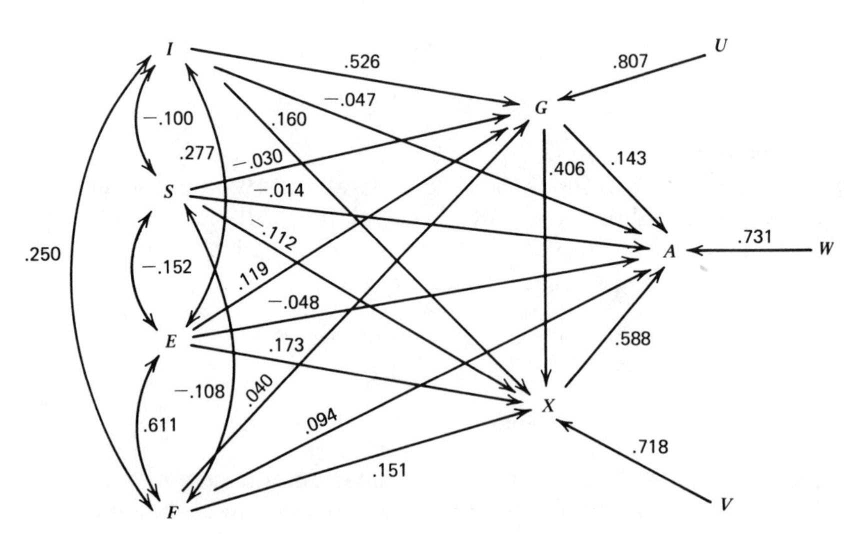

Figure 4.8 Model taken from Kerchoff.

Note that there is a direct effect of intelligence on grades but there are other equivocal effects due to the correlation of intelligence with the other exogenous variables. Since there is no causal model for the purely exogenous variables, one does not know whether the remaining part of the correlation of I with G is due to spuriousness, I and G having common causes, or indirect effects. However, for this particular example, indirect effects are rather implausible; for example, it is implausible to argue that son's IQ causes father's education.

If the causal variable is not purely exogenous, then part of the correlation between it and the variable that it causes is due to spuriousness. For instance, one reason why grades and educational expectation are correlated is the fact that they share the common causes of I, S, E, and F. Spuriousness often explains a considerable portion of the correlation between two variables.

It is then totally illegitimate to find the indirect effects by subtracting the direct effect from the correlation between cause and effect. A highly general method is given by Alwin and Hauser (1975) and it involves the use of reduced form. Let us first translate Figure 4.8 into a set of equations:

$$G = .526I - .030S + .119E + .040F + .807U \qquad [4.14]$$

$$X = .160I - .112S + .173E + .151F + .406G + .718V \qquad [4.15]$$

$$A = -.047I - .014S - .048E + .094F + .143G + .588X + .731W \qquad [4.16]$$

First note that there are no indirect effects of G; there are only direct effects. To find the indirect effects of the four purely exogenous variables on X take Equation 4.15 and substitute Equation 4.14 for G:

$$\begin{aligned} X &= .160I - .112S + .173E + .151F \\ &\quad + .406(.526I - .030S + .119E + .040F + .807U) + .718V \\ &= (.160 + (.406)(.526))I + (-.112 + (.406)(-.030))S \\ &\quad + (.173 + (.406)(.119))E + (.151 + (.406)(.040))F \\ &\quad + (.406)(.807)U + .718V \\ &= .374I - .124S + .221E + .167F + .328U + .718V \end{aligned}$$

(One could combine the two disturbance terms to form a single disturbance.) To find the indirect effect take the coefficient from the reduced form and subtract off the direct effect to find the indirect effect. So the indirect effects are .214 for intelligence, −.012 for number of siblings, .048 for father's education, and .016 for father's occupation.

One need not compute the reduced form coefficients by substituting equations; one could have simply regressed X on I, S, E, and F to obtain the reduced form coefficients. When the number of variables is large, this regression procedure is advisable to minimize both computational and rounding errors. However, reduced form is not necessarily needed to find indirect effects. The tracing rule can be used to find the indirect effects. For instance, there is a single indirect effect from I to X through G and it equals $(.526)(.406) = .214$. To find the indirect effect of G on A, note that it can only go through X. The indirect effect is then $(.406)(.588) = .239$.

To find the indirect effect of I, S, E, and F on A solve for the reduced form. First, substitute Equation 4.14 into G of Equation 4.16:

$$\begin{aligned} A &= -.047I - .014S - .048E + .094F \\ &\quad + .143(.526I - .030S + .119E + .040F + .807U) \\ &\quad + .588X + .731W \\ &= (-.047 + (.143)(.526))I + (-.014 + (.143)(-.030))S \\ &\quad + (-.048E + (.143)(.119))E + (.094 + (.143)(.040))F \\ &\quad + (.143)(.807)U + .588X + .731W \\ &= .028I - .018S - .031E + .100F + .115U + .588X + .731W \end{aligned}$$

Now substitute the *reduced form equation* for X into the preceding equation:

$$\begin{aligned} &= .028I - .018S - .031E + .100F + .115U \\ &\quad + .588(.374I - .124S + .221E + .167F + .328U \\ &\quad + .718V) + .731W \\ &= (.028 + (.588)(.374))I + (-.018 + (.588)(-.124))S \\ &\quad + (-.031 + (.588)(.221))E + (.100 + (.588)(.167))F \\ &\quad + (.115 + (.588)(.328))U + .588(.718)V + .731W \\ &= .248I - .091S + .099E + .198F + .308U + .422V \\ &\quad + .731W \end{aligned} \qquad [4.17]$$

To compute the indirect effect, subtract the direct effects given in Equation 4.16 from the total effects of Equation 4.17. They are then .295 for I, $-.077$ for S, .147 for E, and .104 for F. As a check compute the indirect effects of I on A by the tracing rule. There are three indirect effects: (a) through G, (b) through X, and (c) through G then X. These values are $(.526)(.143) = .075$, $(.160)(.588) = .094$, and $(.526)(.406)(.588) = .126$, respectively. These three values sum to .295, the same value that was obtained from the reduced form method.

If the reduced form coefficients are computed by hand, there is a straightforward way to check the computations. One checks to see if the reduced form equation has a variance of one. This is done here for the reduced form equation of A. Since the disturbances are uncorrelated with each other, the variance of the sum of the disturbances is

$$V(.308V + .422U + .731W) = .308^2 + .422^2 + .731^2$$
$$= .807 \qquad [4.18]$$

To find the variance of the remainder of the equation, simply compute the multiple correlation squared of the reduced form equation:

$$p_{AI}r_{AI} + p_{AS}r_{AS} + p_{AE}r_{AE} + p_{AF}r_{AF}$$
$$= (.248)(.335) + (-.091)(-.153) + (.099)(.303) + (.198)(.331)$$
$$= .193$$

Since the exogenous variables are uncorrelated with the disturbance, the variance of A equals $.807 + .193 = 1$; thus the reduced form coefficients check out.

CONCLUSION

Regression analysis is the backbone of structural modeling. Beta weights estimate path coefficients if the exogenous and endogenous variables are measured and the disturbance is uncorrelated with the exogenous variables. Partial correlation estimates the correlation between disturbances when neither endogenous variable causes the other. In the next chapter the problem of the disturbance correlated with an exogenous variable is considered.

5

Measurement Error in the Exogenous Variable and Third Variables

In the previous chapter it is assumed that the exogenous variables in a structural equation are measured without error and are uncorrelated with the disturbance. In this chapter violations of these two assumptions are considered. It is demonstrated that an estimation procedure called two-stage least squares can be used to estimate such models in some very special circumstances.

The first part of this chapter carefully explores the assumptions of classical measurement theory and recasts them into a path analytic formulation. The second section analytically demonstrates the effect of measurement error in structural equations. The third section illustrates how by adding instrumental variables, measurement error in an exogenous variable can be allowed. The fourth section considers a general estimation procedure for models with instrumental variables called two-stage least squares. The final section outlines the instrumental variable approach to correcting for bias due to unmeasured third variables that cause both the endogenous and exogenous variables.

MEASUREMENT ERROR

What sets aside psychometrics from other statistical approaches is the almost excessive concern about measurement error. Traditionally, measurement error had received only cursory coverage in econometrics. This has since changed and the *errors in variable* problem, measurement error, has been recently given much more attention. However, psychometrics from its inception has highlighted the problem of measurement error. What follows is a discussion of psychometric models of measurement error using path analysis.

The classical model states that an individual's measured score equals the true score plus error of measurement

$$X = T + E \qquad [5.1]$$

where X is measured score, T is true score, and E is error of measurement. It is usually also assumed that the mean of E is zero and the $C(T,E)$ equals zero. The reliability of measure X is defined as

$$\frac{V(T)}{V(X)} \qquad [5.2]$$

that is, the ratio of true variance in the measure to the total variance. Reliability then can be thought of as the percent of variance of the measure that is due to the true score. In this text the usual symbol for reliability of measure X is ρ_{XX} since as is later shown, reliability can be viewed as an "autocorrelation." If the reliability is known, it is often useful to compute the error variance. It equals

$$\begin{aligned} V(E) &= V(X) - V(T) \\ &= V(X)\left[1 - \frac{V(T)}{V(X)}\right] \\ &= V(X)(1 - \rho_{XX}) \qquad [5.3] \end{aligned}$$

The correlation of X with T simply equals

$$\begin{aligned} \rho_{XT} &= \frac{C(X,T)}{(V(X)\,V(T))^{1/2}} \\ &= \frac{V(T)}{(V(X)\,V(T))^{1/2}} \\ &= \left[\frac{V(T)}{V(X)}\right]^{1/2} \\ &= \rho_{XX}{}^{1/2} \qquad [5.4] \end{aligned}$$

Thus the square root of the reliability equals the correlation of the measured score with the true score. The square root of a measure's reliability is the upper limit for the correlation of the measure with any other measure when the errors of measurement of a variable are uncorrelated with all other variables.

Returning to Equation 5.1 it is clear that the variables are not standardized. Note that if $V(X) = 1$, neither T nor E can be standardized since $V(T) + V(E) = 1$. Thus the classical model cannot be directly cast into a path analysis framework. To do so the measurement model must be reformulated as

$$X = aT + bE \qquad [5.5]$$

where X, T, and E are all now *standardized* variables. In Figure 5.1, there is a path diagram for Equation 5.5. From the path diagram it follows that

$$a^2 + b^2 = 1$$

and, therefore,

$$b = (1 - a^2)^{1/2}$$

It also follows that

$$\rho_{TX} = a$$

and

$$\rho_{XX} = a^2$$

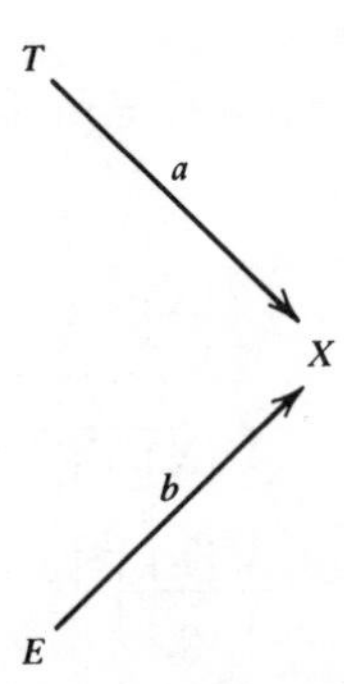

Figure 5.1 Path model for the classical measurement model.

Thus with standardized variables the reliability is simply the square of the path from true score to the measure. The path for the error equals the square root of the quantity one minus the reliability.

There are then two different ways to formulate the measurement model of a single variable: the classical approach of Equation 5.1 and the path analytic approach of Equation 5.5. The text focuses on the path analytic approach, but it is often compared to the classical approach.

In Figure 5.2, two variables, T_X and T_Y, are assumed to be correlated. However, neither variable is measured directly but each with error. The measure X is the errorful indicator of T_X and Y of T_Y. The errors of measurement E_X and E_Y are assumed to be uncorrelated with the true scores and with each other. By the tracing rule it follows that

$$\rho_{XY} = ab\rho_{T_XT_Y}$$

Therefore the correlation between true X and true Y equals

$$\rho_{T_XT_Y} = \frac{\rho_{XY}}{(\rho_{XX}\rho_{YY})^{1/2}}$$

since $a^2 = \rho_{XX}$ and $b^2 = \rho_{YY}$. The preceding formula is the classical correction for attenuation formula. It states that the correlation between true scores equals the correlation between the measured variables divided by the geometric mean of the reliabilities of the measures. The correlation between the measured variables is said to be *attenuated*, and to be equal to the true correlation times some number that is less than or equal to one. The effect, then, of measurement error is to make

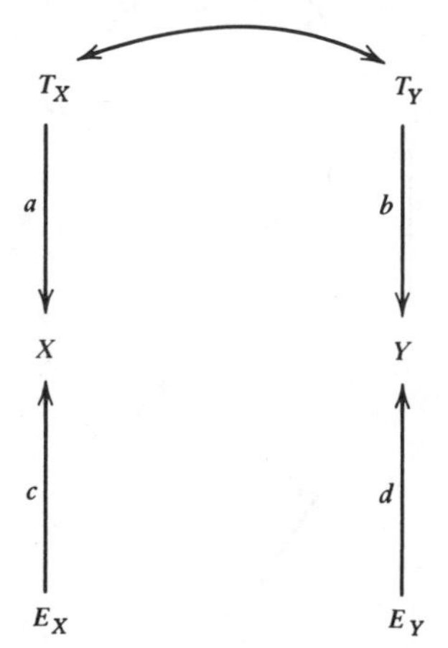

Figure 5.2 Correction for attenuation.

correlations between measures underestimate the absolute value of the true correlation.

Since correlations are biased by measurement error, it should be obvious that ignoring measurement error in variables may seriously distort estimates of structural parameters. Later in this chapter, bias in structural parameters is discussed. Various classical approaches to the estimation of measurement error are considered now.

Typically, estimation of reliability proceeds through the use of other measures. We are all probably familiar with the use of alternative forms or test–retest to estimate reliability. In some sense the logic of these procedures can be seen by examining the path diagram in Figure 5.3. True score T_X causes both X_1 and X_2. Their errors of measurement are mutually uncorrelated and uncorrelated with the true score. It is also specified that X_1 and X_2 are equally reliable ($\rho_{X_1X_1} = \rho_{X_2X_2} = a^2$). Note that it then follows that given uncorrelated errors of measurement the correlation between two equally reliable indicators of the same construct estimates the reliability of the measures. In fact such a correlation is sometimes given as the very definition of reliability.

Classical test theory distinguishes between three different types of measures: parallel, tau equivalent, and congeneric. In this text variables are ordinarily considered in mean deviation form and so only the assumptions made about the equivalence of variances need be considered. Let two measures, X_1 and X_2, have true scores whose correlation is one such that

$$X_1 = T + E_1$$

$$X_2 = kT + E_2$$

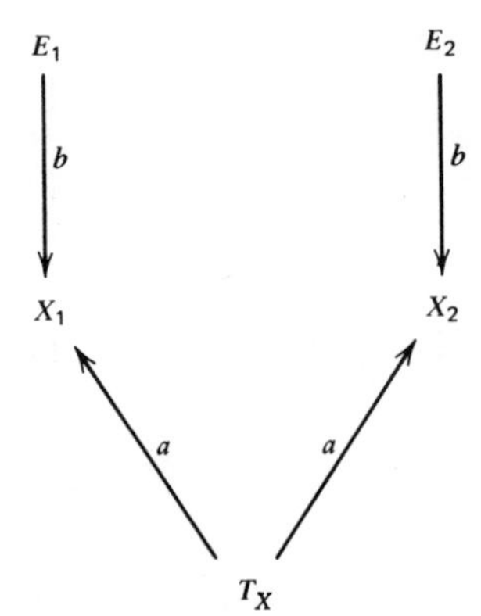

Figure 5.3 Multiple measures model.

where $C(E_1,E_2) = C(E_1,T) = C(E_2,T) = 0$. Measures X_1 and X_2 are said to be *parallel* if $k = 1$ and $V(E_1) = V(E_2)$. It then follows that parallel measures have equal variances and equal covariances; that is, if X_1, X_2, and X_3 are parallel measures, it follows that $V(X_1) = V(X_2) = V(X_3)$ and $C(X_1, X_2) = C(X_1, X_3) = C(X_2, X_3) = V(T)$.

Tests X_1 and X_2 are said to be *tau equivalent* if $k = 1$. Unlike parallel measures, tau-equivalent measures may have unequal error variances. However, the covariance between tau-equivalent measures is constant; that is, if X_1, X_2, and X_3 are tau equivalent then $C(X_1, X_2) = C(X_1, X_3) = C(X_2, X_3) = V(T)$.

Congeneric measures are the most general. Error variances may be unequal and the parameter k need not equal one. The only constraint on the covariances is that they are single factored, which is extensively discussed in Chapter 7.

Different methods of estimating reliability make different assumptions about the measures. For instance, the simple correlation between two measures (assuming uncorrelated errors of measurement) estimates the reliability of only parallel measures. A test–retest correlation also estimates the reliability of parallel measures given no true change. Cronbach's alpha estimates the reliability of the sum of the measures only if the measures are tau-equivalent or parallel. If the measures are congeneric then it underestimates the reliability of the sum.

Standardizing measures may alter their character. Parallel measures remain parallel after standardization since their variances are equal. However, standardization turns tau-equivalent measures into congeneric measures. Ordinarily congeneric measures remain congeneric after standardization with the interesting exception that they become parallel if the original measures were all equally reliable.

Both parallel and tau-equivalent specifications imply that the true score is in the units of measurement of the measured variable. For both cases the true score has a structural coefficient of one. These assumptions then are harmonious with the previously designated classical model of Equation 5.1. Congeneric measures are consistent with the path analytic assumptions of Equation 5.5 since with congeneric measures the true score need not be in the metric of the measure.

MEASUREMENT ERROR IN STRUCTURAL EQUATIONS

The effect of measurement error on the path coefficients is different for the endogenous and exogenous variable. First error in the endogenous variable is considered and then error in the exogenous variable is examined.

Assume two exogenous variables, Z and X, cause Y_T:

$$Y_T = aZ + bX + cU$$

where the disturbance U is assumed to be uncorrelated with Z and X. Both Z and X are measured without error, but Y_T is fallibly measured by Y:

$$Y = Y_T + E$$

where the covariances of E with Y_T, Z, X, and U are zero. If Y, Y_T, and E are standardized then the equation for Y is

$$Y = dY_T + eE$$

where d^2 is the reliability of Y. The unstandardized regression coefficients $b_{YZ \cdot X}$ and $b_{YX \cdot Z}$ yield a and b, respectively. They are unbiased since measurement error in Y is absorbed into its disturbance. However, if the variables are standardized the *beta weights* equal ad and bd. Since d is less than or equal to one, these beta weights will be less than or equal to the absolute value of the true structural coefficient. To summarize, under the classical specification, error of measurement in the endogenous variable does not bias the unstandardized coefficient but it does attenuate the standardized coefficient. To correct the beta weight for attenuation one should divide it by the square root of the reliability of the endogenous variable. Note that if the true beta weight is zero no bias results.

Unfortunately measurement error in the exogenous variable is not nearly so simple. Consider first the following structural equation

$$Y = aX_T + bU$$

where $\rho_{UX_T} = 0$. The variable Y is measured without error but the structural equation for X is

$$X = cX_T + dE$$

where E is uncorrelated with X_T and U. It is assumed that all variables are standardized which makes c^2 the reliability of X. The path from X to Y is then ac. Thus there is attenuation since the true path is a. Even if one assumes the classical specification, the path from X_T to Y is attenuated by c, the square root of the reliability of X.

Consider now two exogenous variables, X_T and Z and an endogenous variable Y. As in Figure 5.4, X_T is measured with error by X. Assuming all the variables in the figure are standardized, the correlations are

$$\rho_{XY} = d(a + fb)$$

$$\rho_{XZ} = df$$

$$\rho_{YZ} = b + af$$

The "path coefficient" from X to Y is then

$$\beta_{YX \cdot Z} = \frac{\rho_{YX} - \rho_{YZ}\,\rho_{XZ}}{1 - \rho_{XZ^2}}$$

$$= \frac{da + dbf - (df)(b + af)}{1 - d^2 f^2}$$

$$= \frac{da\,(1 - f^2)}{1 - d^2 f^2} \qquad [5.6]$$

It can be shown that $d(1 - f^2)/(1 - d^2 f^2)$ equals $\beta_{X_T X \cdot Z}$ or d_Z^2/d where d_Z^2 equals the reliability of X after Z has been partialled out. If f^2 is greater than zero, then d_Z^2/d is less than d. Thus the bias in Equation 5.6 is an attenuation bias. Its size depends mainly on the reliability of X.

The beta coefficient of Y on Z controlling for X equals

$$\beta_{YZ \cdot X} = b + \frac{af - (df)(d(a + bf))}{1 - d^2 f^2}$$

$$= b + \frac{af(1 - d^2)}{1 - d^2 f^2}$$

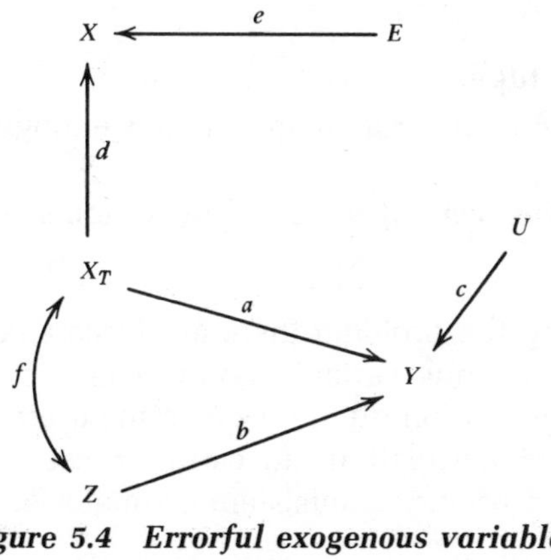

Figure 5.4 Errorful exogenous variable.

Thus the beta weight estimates the true path plus a biasing term. That term equals zero if any of the following three conditions hold:

a. The path from X_T to Y is zero ($a = 0$),
b. the reliability of X is one ($d^2 = 1$), or
c. the correlation of X_T with Z is zero ($f = 0$).

Since X_T is measured with error, one cannot successfully partial out its effects. This can be more clearly seen if X is perfectly unreliable. It then makes no sense to say that we control for X_T by X. If X_T and Z are correlated, then Z will be correlated with the disturbance of Y that contains Y_T.

The direction of the bias is not known in practice. If, however, both a and f are positive, the direction of bias is positive. Thus the bias tends to *overestimate* the path coefficients of the perfectly measured variables, given that the effect of X_T is positive and its correlation with Z is also positive.

If two or more exogenous variables are measured with error, the effects of measurement error are difficult to predict. There are two sources of bias: First, there is attenuation bias due to error in the exogenous variable. Second, added to the attenuated estimate of the coefficient is a term whose size is influenced by the unreliability of the other exogenous variables. It can occur that these two sources of bias cancel each other, but it is unwise to hope for such an outcome.

One must always carefully consider the problem of measurement error in the exogenous variable. Do not assume that its effect will only attenuate estimates of coefficients; it can make the estimate of a truly zero coefficient nonzero or it can yield a coefficient with the opposite sign as the true coefficient. However, a careful analysis may reveal that bias is negligible if

a. Reliability is high.
b. The paths from the true scores to the endogenous variable are small.
c. The exogenous variables have low intercorrelations with each other.

Besides ignoring the problem there are three analytic strategies to the unreliable exogenous variable issue. The first is to correct the correlations for attenuation due to measurement error and then to use these disattenuated correlations to estimate causal parameters. Although this method provides consistent estimates of parameters, it has a shortcoming. It divorces reliability estimation from causal modeling. It has been shown that the measurement process can be expressed in terms of a causal model and therefore all the parameters including reliabilities should be considered as part of one model. This is not the case when one simply disattenuates the correlations. Moreover the resulting parameter estimates cannot be tested for statistical significance in the usual way.

The second strategy for handling error in the exogenous variables is to have parallel forms or multiple indicators of the same construct, which is pursued in Chapters 7 through 9. This strategy introduces unmeasured variables to the model. The third strategy is the method of instrumental variables, which is considered in the next section of this chapter.

INSTRUMENTAL VARIABLES

Consider the path diagram in Figure 5.5. If there were a path from Y to Z with X, Y, and Z measured, the model would be underidentified since there are four free parameters and only three correlations. The instrumental variable strategy is to buy the identification of the reliability coefficient by assuming a path from an exogenous variable is zero. In Figure 5.5 there are only three free parameters since there is no path from Y to Z. Note that since $\rho_{YZ} = a\rho_{X_TY}$ and $\rho_{X_TY} = \rho_{XY}/b$

$$\frac{a}{b} = \frac{\rho_{YZ}}{\rho_{XY}}$$

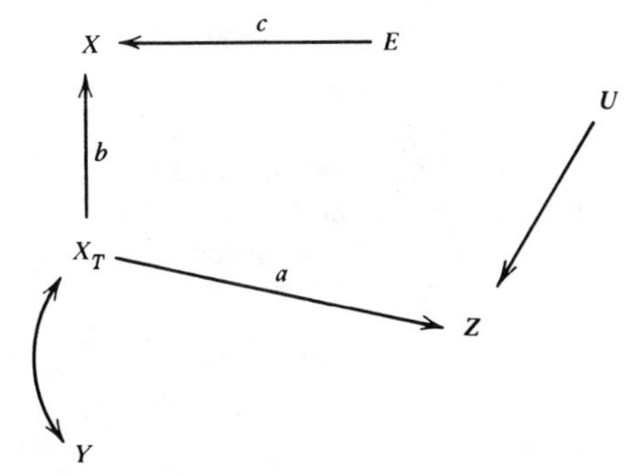

Figure 5.5 Instrumental variable illustration.

Moreover, by the tracing rule $\rho_{XZ} = ab$. It then follows that

$$b^2 = \frac{\rho_{XZ}\rho_{XY}}{\rho_{YZ}}$$

and

$$a^2 = \frac{\rho_{XZ}\rho_{YZ}}{\rho_{XY}}$$

Since b is usually taken to be positive, the sign of a is determined by ρ_{XZ} or ρ_{YZ}/ρ_{XY} both of which should have the same sign.

If one takes the classical specification of test score theory ($X = X_T + E$), then

$$a = \frac{C(Y,Z)}{C(X,Y)}$$

and

$$\rho_{XX} = \frac{\rho_{XZ}\rho_{XY}}{\rho_{YZ}}$$

Clearly to be able to estimate a both ρ_{XY} and b^2 should not be small; otherwise, empirical underidentification results.

Instrumental variable estimation can disasterously break down if the model is misspecified. Consider the set of correlations taken from unpublished data of L. Brush in Table 5.1. The model specified in Figure 5.6 is given for illustrative purposes only and was not specified by Brush. Sex S does not affect expectations E, but rather its effect is mediated through anxiety X and ability A. The anxiety variable is assumed to be measured with error. Using the first law

$$\rho_{ES} = a\rho_{SX_T} + b\rho_{SA}$$

$$\rho_{EA} = b + a\rho_{AX_T}$$

Table 5.1. Correlations of Sex, Ability, Anxiety, and Expectations for Success in Mathematics

	S	A	X	E
S	1.00			
A	.21	1.00		
X	.14	.70	1.00	
E	.21	.62	.53	1.00

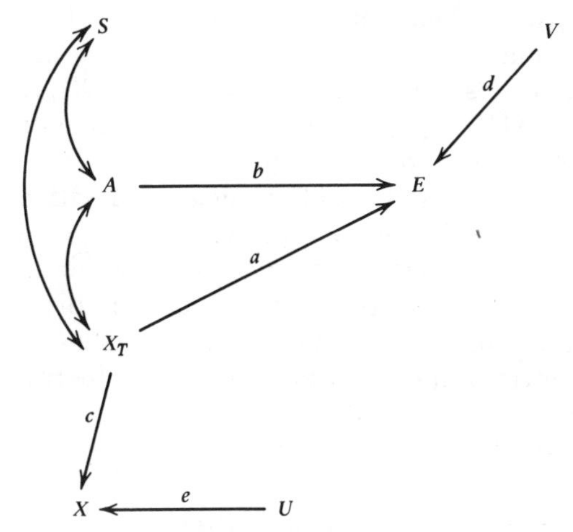

Figure 5.6 Breakdown of instrumental variable estimation.

Since $\rho_{AX_T} = \rho_{AX}/c$ and $\rho_{SX_T} = \dfrac{\rho_{SX}}{c}$,

$$\rho_{ES} = \frac{\rho_{SX}a}{c} + b\rho_{SA}$$

$$\rho_{EA} = b + \frac{\rho_{AX}a}{c}$$

The solution for a/c is

$$\frac{a}{c} = \frac{\rho_{ES} - \rho_{SA}\,\rho_{EA}}{\rho_{SX} - \rho_{AX}\,\rho_{SA}}$$

and

$$b = \frac{\rho_{SX}\,\rho_{EA} - \rho_{AX}\,\rho_{ES}}{\rho_{SX} - \rho_{AX}\,\rho_{SA}}$$

The estimate for a/c is -11.40 and for b is 8.60 using the estimated correlations in Table 5.1. Such results, though not impossible, are highly implausible. If one proceeds to solve for a, ρ_{AX_T}, and d further implausible values arise. These highly anomalous values are possible if instrumental variable estimation is carelessly applied.

Instrumental variable estimation can yield informative estimates as in the next example taken from Jöreskog (1973). Children were measured at fifth (X_1 and X_2), seventh (X_3 and X_4), and ninth grade (X_5 and X_6). Of interest are tests of mathematical ability (X_1, X_3, and X_5) and science knowledge (X_2, X_4, and X_6). The correlation matrix is contained in Table 5.2 and the model in Figure 5.7. The two seventh grade measures (X_3 and X_4) are assumed to be measured with error. The wave 1 measures are taken to be instruments and the wave 3 measures as the endogenous variables. Clearly there is measurement error in these variables also, but errors in the endogenous variables only attenuate estimates and error in the instruments only reduces the efficiency of estimation. However, it must be assumed that such errors are mutually uncorrelated.

It can be shown that

$$\rho_{15} = \frac{\rho_{13}a}{e} + \frac{\rho_{14}b}{f}$$

$$\rho_{25} = \frac{\rho_{23}a}{e} + \frac{\rho_{24}b}{f}$$

$$\rho_{16} = \frac{\rho_{13}c}{e} + \frac{\rho_{14}d}{f}$$

$$\rho_{26} = \frac{\rho_{23}c}{e} + \frac{\rho_{24}d}{f}$$

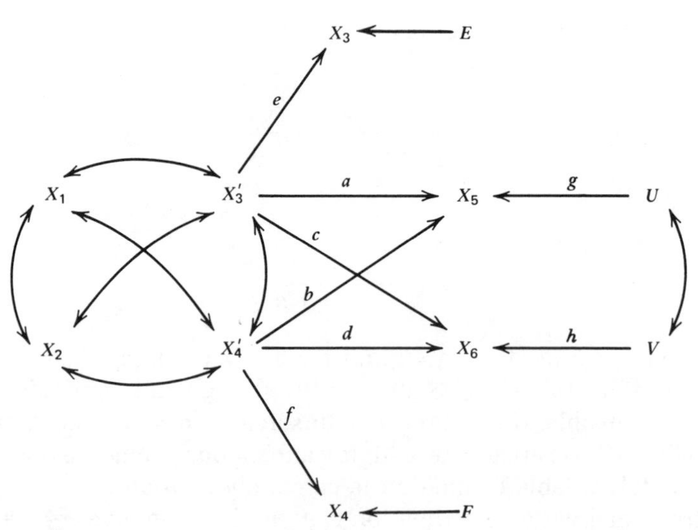

Figure 5.7 Mathematics and science ability example.

The solutions then are

$$\frac{a}{e} = \frac{\rho_{15}\,\rho_{24} - \rho_{14}\,\rho_{25}}{\rho_{13}\,\rho_{24} - \rho_{23}\,\rho_{14}}$$

$$\frac{b}{f} = \frac{\rho_{13}\,\rho_{25} - \rho_{23}\,\rho_{15}}{\rho_{13}\,\rho_{24} - \rho_{23}\,\rho_{14}}$$

$$\frac{c}{e} = \frac{\rho_{16}\,\rho_{24} - \rho_{14}\,\rho_{26}}{\rho_{13}\,\rho_{24} - \rho_{23}\,\rho_{14}}$$

$$\frac{d}{f} = \frac{\rho_{13}\,\rho_{26} - \rho_{23}\,\rho_{16}}{\rho_{13}\,\rho_{24} - \rho_{23}\,\rho_{14}}$$

Substituting in the correlations from Table 5.2 yields

$$\frac{a}{e}\text{: } 1.071$$

$$\frac{b}{f}\text{: } -.108$$

$$\frac{c}{e}\text{: } -.144$$

$$\frac{d}{f}\text{: } 1.112$$

It also follows that

$$ae = \rho_{35} - \frac{\rho_{34}b}{f}$$

$$bf = \rho_{45} - \frac{\rho_{34}a}{e}$$

Table 5.2. Mathematics and Science Ability Measured at Fifth, Seventh, and Ninth Grades[a,b]

X_1	1.000					
X_2	.755	1.000				
X_3	.730	.664	1.000			
X_4	.651	.700	.710	1.000		
X_5	.712	.636	.763	.673	1.000	
X_6	.619	.683	.671	.740	.736	1.000
	X_1	X_2	X_3	X_4	X_5	X_6

[a]Data taken from Jöreskog (1973).
[b]N = 730.

$$ce = \rho_{36} - \frac{\rho_{34}d}{f}$$

$$df = \rho_{46} - \frac{\rho_{34}c}{e}$$

yielding

$$\begin{aligned} ae&: \ \ .840 \\ bf&: -.088 \\ ce&: -.119 \\ df&: \ \ .842 \end{aligned}$$

It then follows that

$$\begin{aligned} a&: \ \ .948 \\ b&: -.097 \\ c&: -.131 \\ d&: \ \ .968 \\ e^2&: \ \ .784,\ .824 \\ f^2&: \ \ .814,\ .757 \end{aligned}$$

The two negative estimates seem somewhat anomalous but perhaps proficiency in one area interferes with ability in other areas. Compare the ordinary multiple regression estimates with the preceding and note how markedly different they are:

$$\begin{aligned} a&: .575 \\ b&: .265 \\ c&: .294 \\ d&: .532 \end{aligned}$$

Since there are two estimates of e^2 and f^2, a simple pooling strategy is taken by using an arithmetic average. The estimate of $\rho_{3'4'}$ is then $.71/((.804)(.786))^{1/2} = .893$. The estimates of g and h are

$$g = (1 - a^2 - b^2 - 2ab\rho_{3'4'})^{1/2} = .508$$
$$h = (1 - c^2 - d^2 - 2cd\rho_{3'4'})^{1/2} = .522$$

and ρ_{UV} is

$$\frac{(\rho_{56} - ad\rho_{3'4'} - bc\rho_{3'4'} - ac - bd)}{gh} = .463$$

The instrumental variable estimates yield a set of generally plausible and informative values.

TWO-STAGE LEAST SQUARES ESTIMATION

The classical method of estimating parameters from models with instrumental variables is called two-stage least squares (2SLS). This section first explains the estimation procedure and then discusses the assumptions necessary for such an estimation procedure.

Denote the endogenous variable as Z and its disturbance as U. The set of n variables which are assumed to be measured with error are called X; the set of p instruments are called I; and the set of q exogenous variables measured without error are called Y. The structural equation then is

$$Z = \overset{p}{\Sigma} a_i X_i' + \overset{q}{\Sigma} b_k Y_k + U$$

where the prime designates X_i true. Considered will be two alternative specifications of the measurement model. First the classical model where

$$X_i = X_i' + E_i$$

with $V(X_i')/V(X_i) = c_i^2$. The path analytic specification is

$$X_i = c_i X_i' + (1 - c_i^2)^{1/2} E_i$$

where $V(X_i) = V(X_i') = V(E_i)$. Note this second specification does not assume that the X variables are standardized but only that measured variance equals true variance. For both specifications the reliability of X_i is c_i^2.

As its name suggests two-stage least squares involves two different multiple regression runs. The first stage involves regressing each X_i on I and Y. This will involve n separate regression equations. From each regression one can estimate $\hat{X}_i$, that is, the predicted X_i given I and Y. Note that if X_i is standardized, $\hat{X}_i$ will not ordinarily have unit variance. The results of this first stage of estimation can be thought of as an attempt to estimate the true scores of the X_i. Since it will be seen that the measurement error in X_i is uncorrelated with I and Y, the $\hat{X}_i$ contain only true score variance.

Now that the X variables have been purified the second stage can begin. The endogenous variable Z is now regressed on $\hat{X}$ and Y. The interpretation of the coefficients for the Y variables is straightforward. They are the estimates of b_k, the structural coefficients for Y_k. The interpretation of the coefficients for the $\hat{X}$ depends on the test theory model that is chosen. If the classical model is chosen, then they are simply the structural coefficients for the true scores. If the path analytic model is chosen another stage of estimation remains. It can be shown that second stage estimates equal a_i/c_i. It also can be shown that

$$a_ic_i = \frac{C(X_i,Z) - \sum_{m\neq i} C(X_i,X_m)\, a_m/c_m - \sum_k C(X_i,Y_k)b_k}{V(X_i)}$$

Since all the preceding terms on the left-hand side can be estimated from the data (recall that a_m/c_m are the second stage estimates for X_m), one can solve for an estimate a_ic_i. Then with estimates of a_i/c_i and a_ic_i, one can solve for a_i and c_i^2. Note that the estimates of c_i^2 should, in principle, be less than one. Also for a_i not to be imaginary, the solution for a_ic_i and a_i/c_i should both have the same sign.

To estimate the reliability of X_i given the classical test theory specification compute

$$c_i^2 = \frac{C(X_i,Z) - \sum_{m\neq i} C(X_i,X_m)\, a_m - \sum_k C(X_i,Y_k)b_k}{V(X_i)\, a_i}$$

where a_i, a_m, and b_i are the second stage coefficients.

The second stage coefficients can be tested for significance in the usual way. Note, however, that $\hat{X}$ and Y variables may be highly colinear since the Y variables were used to estimate $\hat{X}$. The power of the tests of significance may then be rather low.

If there are more instruments than variables needing instruments ($p > n$), the model is overidentified. One advantage of two-stage least squares is that overidentified models present no special problem. For complicated models, however, it is not a simple matter to test the overidentifying restriction. One strategy is as follows: Regress each I on $\hat{X}$ and Y; compute for each I_j the residual $I_j - \hat{I}_j$; and denote them as Q. Although there were originally n instrumental variables, only p minus n of the Q variables are independent. Now compute the first $p - n$ principal components of Q. Now include these $p - n$ components in the second stage regression equation. The test of the overidentifying restrictions is simply whether these principal components' coefficients are zero.

To apply instrumental variable estimation the following assumptions must hold:

1. The errors of measurement in the X variables must be uncorrelated with each other and the other variables in the model.
2. The instruments must not directly cause the endogenous variable.
3. The instruments must be uncorrelated with the disturbance in the endogenous variable.
4. The number of variables needing instruments must be less than or equal to the number of instruments.

The first assumption is a classical test theory specification. The second assumption creates the zero paths which bring about identification. In a sense this is the key assumption. The instrument can cause Z but its effect must be mediated by X' or Y. The third assumption is the usual uncorrelated disturbance assumption. The fourth assumption is self-explanatory.

To avoid empirical underidentification one must assume that the I and X' variables share variance independent of the Y variables. For instance, X' may cause I or vice versa, or some third variable besides Y may cause them. If there is little or no independent association between X' and I, the $\hat{X}$ will be virtually entirely explained by the Y variables. Then on the second stage the $\hat{X}$ and Y variables will be sufficiently colinear to preclude reliable estimation.

Secondly, the estimates of instrumental variable regression will be empirically underidentified if weights for the I variables on two or more X variables are proportional. For instance, if

$$\hat{X}_1 = .2I_1 + .3I_2$$

$$\hat{X}_2 = .4I_1 + .6I_2$$

then $\rho_{\hat{X}_1\hat{X}_2} = 1$ resulting in perfect multicolinearity between $\hat{X}_1$ and $\hat{X}_2$ in the second stage. If the coefficients of the I variables for $\hat{X}_j$ are linearly or are nearly linearly dependent with those of any other set of other $\hat{X}$ variables, empirical underidentification results.

One advantage of a two-stage least-squares solution is that it can estimate the parameters of an overidentified model. For some simple models the test of the overidentifying restrictions is rather straightforward. Imagine the case in which X_1 causes X_4 and X_1 is measured with error and both X_2 and X_3 can be used as instruments of X_1. Since both r_{24}/r_{12} and r_{34}/r_{13} estimate the same parameter, there is the following overidentifying restriction:

$$\rho_{24}\rho_{13} - \rho_{34}\rho_{12} = 0$$

which is called a vanishing tetrad; its significance test is discussed in Chapter 7.

Although Duncan (1975, p. 98) has argued that the test of the overidentifying restriction in this instance is of little practical value, such a test of the overidentifying restriction should be performed. Take first the case in which the overidentifying restriction is met; that is, the correlations satisfy the overidentifying restriction within the limits of sampling error. Duncan claims that this is of little value since it is a test of the null hypothesis. One should not, according to Duncan, accept that the overidentifying restriction holds, but rather one cannot accept that it does not hold. I do not accept this conclusion. We often find ourselves testing the null hypothesis. In this case, the test of the null hypothesis tells us whether we can safely conclude that either the instrument does not cause the exogenous variable or is correlated with the disturbance. Since many of us have difficulty of thinking of good instrumental variables, the overidentifying restriction should be helpful in that it allows us to test our assumptions. As with any significance test, sample size should be large and power high. If one stringently objects to testing the null hypothesis, one could set up a reasonable alternative hypothesis and treat it as the null hypothesis.

Duncan is just as forceful in arguing against the usefulness of an overidentifying restriction not being met. For instance, in the case preceding, if the overidentification is not met one does not know whether X_2 or X_3, or both, are bad instruments. If an overidentifying restriction is not met, it rarely localizes the source of specification error. However, not meeting the overidentifying restriction clearly tells us that something is amiss. It forces us to rethink and respecify the model. If the model is just-identified, we are always lulled into the feeling that the model is correctly specified. Overidentification affords us with many opportunities to be humbled. It does not bring truth or certainty, but like any good empirical test, it offers the possibility of confirmation or rejection of theory. If the model is overidentified the researcher should make every effort to test the overidentifying restrictions.

UNMEASURED THIRD VARIABLE

Instrumental variable estimation can also be applied to the case in which an exogenous variable is correlated with the disturbance of the endogenous variable. This would happen if there were an unmeasured third variable that caused both the exogenous and endogenous variables. There must be as many instrumental variables as there are exogenous variables that are correlated with the disturbance. Again the instrumental variables by definition cannot cause the endogenous variable or be correlated with the disturbance. There are two important differences for this case from the case in which the exogenous variable is measured with error. First, the instrument should not be caused by the variable that needs an instrument. If it is caused, then the instrument would most probably be correlated with the disturbance of the endogenous variable. Second, when the path-analytic specification is assumed there is no need to go through a further step to solve for the structural parameters.

Consider the model in Figure 5.8. The variables are taken from Table 4.1 of the previous chapter. The variable X is educational aspirations. E is father's education, G is grades, F is father's occupation, and I is intelligence. Both E and G are assumed to be correlated with the disturbance in X and F and I are instruments. To solve for a and b one could employ two-stage least squares. However, one can also solve for the relevant parameters by path analysis. The two relevant equations are

$$\rho_{XF} = a\rho_{FE} + b\rho_{FG}$$

$$\rho_{XI} = a\rho_{IE} + b\rho_{IG}$$

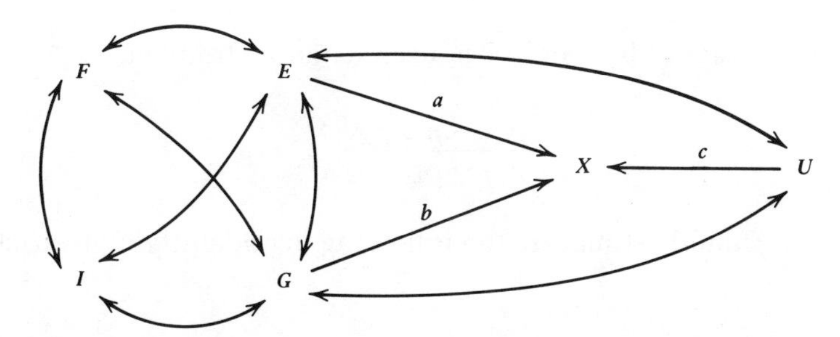

***Figure 5.8** Kerchoff example with two exogenous variables correlated with the disturbance.*

The solution then is

$$a = \frac{\rho_{FX}\,\rho_{IG} - \rho_{IX}\,\rho_{FG}}{\rho_{FE}\,\rho_{IG} - \rho_{FG}\,\rho_{IE}}$$

$$b = \frac{\rho_{IX}\,\rho_{FE} - \rho_{FX}\,\rho_{IE}}{\rho_{FE}\,\rho_{IG} - \rho_{FG}\,\rho_{IE}}$$

Substituting in the correlations in Table 4.1 a value of .403 is obtained for a and .660 for b. The solution for c is

$$V(cU) = V(X - aE - bG)$$

$$c^2 = 1 + a^2 + b^2 + 2ab\rho_{EG} - 2a\rho_{EX} - 2b\rho_{GX}$$

which yields .607 for c^2 and .779 for c. The correlations of the exogenous variable with the disturbance can be solved from the correlation of the exogenous variable with the endogenous variable

$$\rho_{EX} = a + b\rho_{EG} + c\rho_{EU}$$

$$\rho_{GX} = a\rho_{EG} + b + c\rho_{GU}$$

which yields a value of $-.194$ for r_{EU} and $-.233$ for r_{GU}. These negative values seem implausible.

It is instructive to derive the overidentifying restriction that tests whether $\rho_{EU} = 0$. If ρ_{EU} were zero, then both F and I could be used as instruments for G. There is then an excess of one instrument. The estimate of b, if only F is used as an instrument, is

$$\frac{\rho_{FX} - \rho_{FE}\,\rho_{XE}}{\rho_{FG} - \rho_{EG}\,\rho_{FE}}$$

and the estimate of b, if only I is used as an instrument, is

$$\frac{\rho_{IX} - \rho_{IE}\,\rho_{XE}}{\rho_{IG} - \rho_{EG}\,\rho_{IE}}$$

Since both should estimate b, the following overidentifying restriction holds:

$$(\rho_{FX} - \rho_{FE}\,\rho_{XE})(\rho_{IG} - \rho_{EG}\,\rho_{IE}) = (\rho_{IX} - \rho_{IE}\,\rho_{XE})(\rho_{FG} - \rho_{EG}\,\rho_{FE})$$

which implies that

$$\rho_{FX\cdot E}\, \rho_{IG\cdot E} - \rho_{IX\cdot E}\, \rho_{FG\cdot E} = 0$$

which is a vanishing tetrad (see Chapter 7) of partial correlations.

This chapter has discussed a number of topics: The theory of measurement error, error in exogenous variables, and exogenous variable–disturbance correlations. Chapters 7, 8, and 9 return to the problem of measurement error. In the next chapter it is demonstrated that instrumental variable estimation can be applied to nonhierarchical models.

6

Observed Variables As Causes of Each Other

All the models discussed in the previous two chapters have not allowed feedback. Those models did not allow for a variable to both cause and be caused by another variable. In this chapter attention is directed toward nonhierarchical or feedback models, which the econometricians call *simultaneity*. The chapter is divided into three sections. The first section outlines general considerations of nonhierarchical models. The second section discusses estimation. The third and final section presents a nonhierarchical model for halo effect.

GENERAL CONSIDERATIONS

The simplest nonhierarchical model is

$$X_1 = aX_2 + bU \quad [6.1]$$

$$X_2 = cX_1 + dV \quad [6.2]$$

where $\rho_{UV} = 0$, as in the path diagram in Figure 6.1. It is later shown that the disturbances of each equation are correlated with the exogenous variable in the equation, that is, $\rho_{1V} \neq 0$ and $\rho_{2U} \neq 0$. Since each variable causes the other, it is not possible to use multiple regression to estimate the path coefficients. There is another obvious problem with

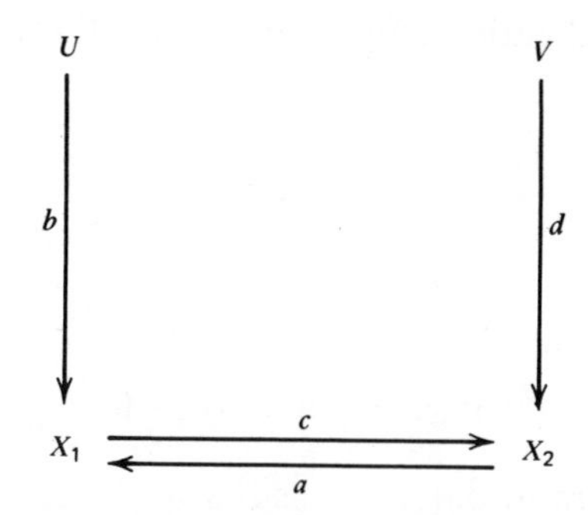

Figure 6.1 A simple feedback model.

the model in Figure 6.1. There are two free parameters, a and c, and only one correlation if only X_1 and X_2 are measured. The model then is underidentified.

To derive the correlation between X_1 and X_2, take the covariance between Equations 6.1 and 6.2 to obtain

$$\rho_{12} = ac\rho_{12} + bc\rho_{1U} + ad\rho_{2V} \qquad [6.3]$$

To solve for ρ_{1U} and ρ_{2V} take the covariance of X_1 with Equation 6.1 and X_2 with 6.2 to obtain

$$1 = a\rho_{12} + b\rho_{1U}$$

and

$$1 = c\rho_{12} + d\rho_{2V}$$

Rearranging yields

$$b\rho_{1U} = 1 - a\rho_{12}$$

and

$$d\rho_{2V} = 1 - c\rho_{12}$$

Substituting these values for $b\rho_{1U}$ and $d\rho_{2V}$ into Equation 6.3 yields

$$\rho_{12} = ac\rho_{12} + c(1-a\rho_{12}) + a(1-c\rho_{12})$$

Simplifying gives the result of

$$\rho_{12} = \frac{a + c}{1 + ac}$$

To achieve identification one might assume that $a = c$ and then solve for a:

$$\rho_{12} = \frac{2a}{1 + a^2}$$

Rearranging yields

$$\rho_{12}a^2 - 2a + \rho_{12} = 0$$

Applying the quadratic theorem results in

$$a = \frac{2 \pm (4 - 4\rho_{12}^2)^{1/2}}{2\rho_{12}}$$

$$= \frac{1 \pm (1 - \rho_{12}^2)^{1/2}}{\rho_{12}}$$

As an example, if $\rho_{12} = .5$ then a and c would both equal .268 or 3.732. (Recall from Chapter 4 that standardized regression coefficients can be greater than one.)

Equilibrium Assumption

For the model in Figure 6.1, assume that $a = .4$ and $c = .6$. It then follows that $\rho_{12} = .806$. If the score of a person on X_1 is increased from zero to one, the person's score on X_2 is changed by .6. But this in turn affects the X_1 score, since X_2 causes X_1. The score on X_1 is now increased by $(.6)(.4) = .24$. This in turn affects X_2 *ad infinitum*. The total change in X_2 is

$$.6 + .6(.6)(.4) + .6(.6)^2(.4)^2 + \cdots + .6(.6)^i(.4)^i + \cdots$$

or

$$\frac{.6}{1-(.6)(.4)}$$

which equals .790.

Since the change in X_2 echoed back to change X_1, the X_1 score changes by more than the initial 1. The total change in the person's score on X_1 is

$$1 + (.6)(.4) + (.6)^2(.4)^2 + \cdots + (.6)^i(.4)^i + \cdots$$

or

$$\frac{1}{1-(.6)(.4)}$$

which equals 1.32. In general, with Equations 6.1 and 6.2 if a person's score on X_1 is changed by g units, the total change in X_2 is $cg/(1-ac)$ and in X_1 is $g/(1-ac)$. Since it was stated in Chapter 1 that causation is ordinarily assumed to occur with a time lag, a model with feedback would create effects that would last forever. Of course the effect may very well dampen out very quickly, but in theory the effect never stops. Because of this "eternal" effect of feedback it is critical to examine systems of variables that are in equilibrium, that is, where a causal effect is not just beginning but has generally dampened out. Heise (1975, p. 227) nicely illustrates how mistaken conclusions can be made when the system of variables is not in equilibrium.

Modified Tracing Rule

Feedback models create very special issues for the analyst. For instance, although the first law still holds, with feedback models the tracing rule does not hold in a simple way. The tracing rule can be modified so that it may hold for feedback models. The discussion must be divided into two parts. First discussed is the case of determining ρ_{12} where X_1 causes X_2 and X_2 and X_3 are involved in a feedback relationship. The key here is that X_1 and X_2 are *not* in a feedback relationship. Examining the path diagram in Figure 6.2, one might be tempted to apply the tracing rule and say ρ_{12} equals a. Examine, however, the tracings from X_1 to X_2. There is first a path directly from X_1 to X_2 but there is also a path from X_1 to X_2 to X_3 and back again to X_2. This would seem to violate the tracing rule since that rule does not allow tracing through the same variable twice. One can, however, argue that since

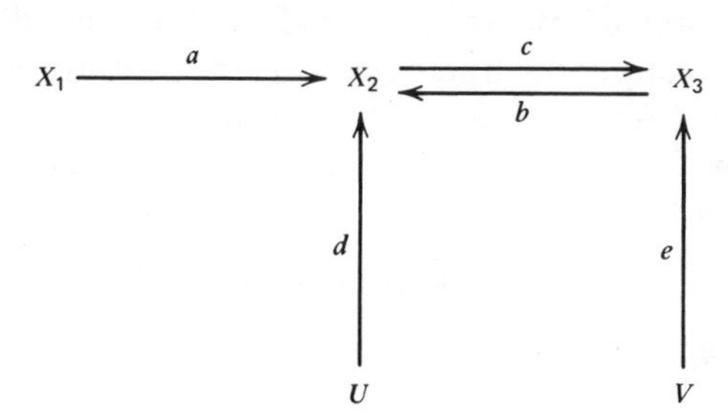

Figure 6.2 An exogenous variable in a nonhierarchical system.

due to feedback the values of X_2 change, the "same" variable is not traced through. Whatever the rationale there are the following paths: $a + abc + a(bc)^2 + a(bc)^3$ and so on. It then follows that

$$\rho_{12} = a \sum_{i=0}^{\infty} (bc)^i = \frac{a}{1-bc}$$

Thus in a sense the tracing rule still applies but there are an infinite number of tracings.

The correlation ρ_{13} in Figure 6.2 can be computed in a similar fashion:

$$\rho_{13} = ac \sum_{i=0}^{\infty} (bc)^i = \frac{ac}{1-bc}$$

If a path is added from X_1 to X_3 and is labeled f it would then follow that

$$\rho_{12} = \frac{a+fb}{1-bc}$$

$$\rho_{13} = \frac{f+ac}{1-bc}$$

Thus to apply the *modified tracing rule* for the correlation between a variable and a second variable that is involved in a feedback relationship with a third variable, simply apply the tracing rule learned in Chapter 3. This quantity is then divided by $1 - gh$, where g is the path from one of the feedback variables to the other and h is the second feedback path.

The second part of the modified rule concerns a correlation between X_1 and X_2, where X_1 is in a feedback relationship and X_2 is directly or indirectly caused by X_1. The feedback relationship that X_1 is in may or may not involve X_2 but X_1 does cause X_2. First find the reduced form of the variable in the feedback relationship, that is, the equation without feedback incorporated into it. Then redraw the path diagram for the reduced form and apply the tracing rule. (This procedure also works for the previous part.) As an example, to determine ρ_{23} in Figure 6.2, write the equation for X_2, $aX_1 + bX_3 + dU$, and for X_3, $cX_2 + eV$. Substituting the equation for X_3 into the X_2 equation yields

$$X_2 = aX_1 + b(cX_2+eV) + dU$$

which reduces to

$$X_2 = \frac{a}{(1-bc)}X_1 + \frac{d}{(1-bc)}U + \frac{be}{(1-bc)}V$$

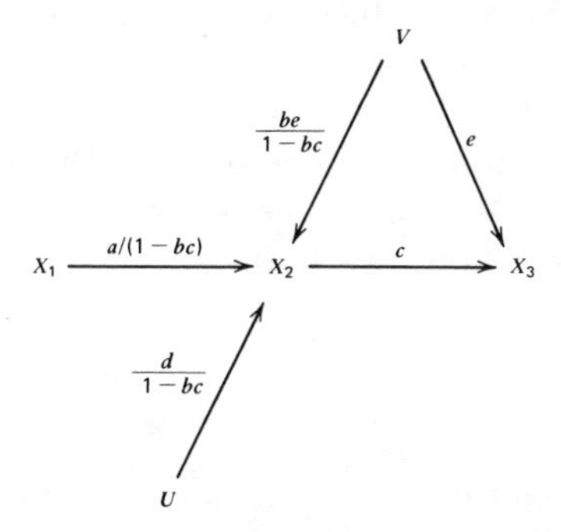

Figure 6.3 Reduced form path diagram.

Now to compute ρ_{23} redraw the path diagram as in Figure 6.3. The correlation between X_2 and X_3 then is $c + be^2/(1-bc)$. Note that a different but equivalent answer would have been obtained had the reduced form of X_3 been solved for. Although the tracing rule is valid when so modified, generally it is much simpler to apply the first law. I state the rule only because it may occasionally be helpful.

Returning to the model in Figure 6.2, it happens to be identified. As has been shown

$$\rho_{12} = \frac{a}{1-bc}$$

$$\rho_{13} = \frac{ac}{1-bc}$$

and with some algebraic work it can be shown that

$$\rho_{23} = \frac{b+c}{1+bc}$$

The solution for c is straightforward:

$$c = \frac{\rho_{13}}{\rho_{12}}$$

Since c is known, one can solve for b from the ρ_{23} equation to find

$$b = \frac{\rho_{13} - \rho_{23}\rho_{12}}{\rho_{23}\rho_{13} - \rho_{12}}$$

and a in terms of b and c

$$a = \rho_{12}(1-bc)$$

To test whether a is zero, one tests whether $\rho_{12} = 0$; to test if b is zero, $\rho_{13\cdot2} = 0$; and to test c we see if $\rho_{13} = 0$. It should be noted that the model in Figure 6.2 is not identified if $\rho_{UV} \neq 0$.

To estimate d take the covariance of X_2 with the equation for X_2 to obtain $1 = a\rho_{12} + b\rho_{23} + d\rho_{2U}$. Since $\rho_{2U} = d/(1-bc)$ by the use of the modified tracing rule, it follows that

$$1 = a\rho_{12} + b\rho_{23} + \frac{d^2}{1-bc}$$

Now solve for d to obtain

$$d = ((1-a\rho_{12}-b\rho_{23})(1-bc))^{1/2}$$

A simpler and more direct solution for d is to take the variance of

$$dU = X_2 - aX_1 - bX_3$$

which is

$$d^2 = 1 + a^2 + b^2 + 2ab\rho_{13} - 2a\rho_{12} - 2b\rho_{23}$$

The solution for d is then the square root of the preceding expression.

By a similar procedure one can obtain e as

$$(1 + c^2 - 2c\rho_{23})^{1/2}$$

or one can derive e by taking the covariance of X_3 with its structural equation

$$e = ((1-c\rho_{23})(1-bc))^{1/2}$$

Needless to say the two solutions for both d and e are algebraically equivalent.

Indirect Feedback

Consider the model in Figure 6.4. There is an indirect feedback loop: X_1 causes X_3 which in turn causes X_2, and X_2 causes X_1. The equations are

$$X_1 = eX_2 + fU$$
$$X_2 = gX_3 + hW$$
$$X_3 = iX_1 + jS \qquad [6.4]$$

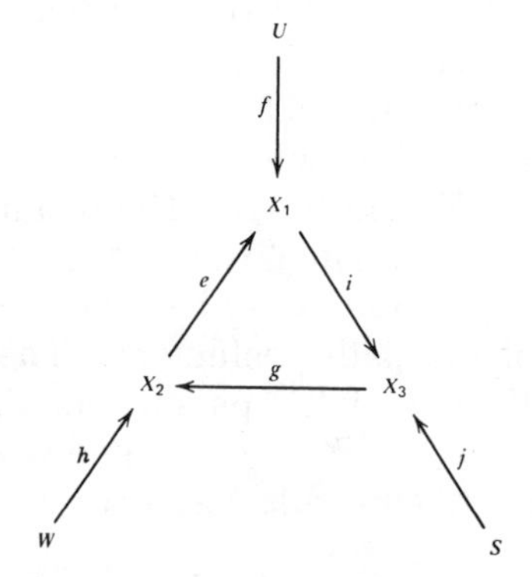

Figure 6.4 Indirect feedback.

Substituting the equation of X_3 into the equation of X_2 yields

$$X_2 = giX_1 + gjS + hW \qquad [6.5]$$

Note that there there now is a formal equivalence between the two equations 6.4 and 6.5 and the two equations 6.1 and 6.2; that is, let $e = a$, $f = b$, $gi = c$, and $gjS + hW = dV$. The correlation then between X_1 and X_2 is

$$\frac{e + gi}{1 + egi}$$

and similarly

$$\rho_{13} = \frac{i + eg}{1 + ieg}$$

and

$$\rho_{23} = \frac{g + ie}{1 + gie}$$

Although the model is just-identified with three free parameters and three correlations, there is no simple solution to the equations.

ESTIMATION

Most of the preceding discussion would not be found in the consideration of feedback models in other texts. Those texts concentrate on the topic in this section: the use of two-stage least squares to estimate parameters for models with feedback.

Consider the model in Figure 6.5. Variables X_3 and X_4 are involved in a feedback relationship. Variables X_1 and X_2 are purely exogenous with X_1 causing X_3 and X_2 causing X_4. The disturbances of X_3 and X_4 have been allowed to be correlated. Naively, one might think that one could simply regress X_3 on both X_1 and X_4 and regress X_4 on X_2 and X_3 to obtain estimates of the path coefficients. They do not ordinarily provide unbiased estimates of the parameters since X_4 is correlated with the disturbance of X_3 and X_3 is correlated with the disturbance of X_4. Using the modified tracing rule these correlations are

$$\rho_{4U} = \frac{ed + f\rho_{UV}}{1 - cd}$$

and

$$\rho_{3V} = \frac{fc + e\rho_{UV}}{1 - cd}$$

Note the correlation still exists even if $\rho_{UV} = 0$. Since the disturbance is correlated with a cause, simple multiple regression breaks down. However, as was seen in the last chapter, in the case of an exogenous variable correlated with a disturbance, identification is still possible given certain zero paths. In Figure 6.5 estimation may be possible since there is no path from X_2 to X_3 and from X_1 to X_4. Using the terminology of the previous chapter, X_2 can be used as an instrument for the X_3-equation and X_1 for the X_4-equation. For the X_3-equation the first law yields

$$\rho_{13} = a + c\rho_{14}$$

$$\rho_{23} = a\rho_{12} + c\rho_{24}$$

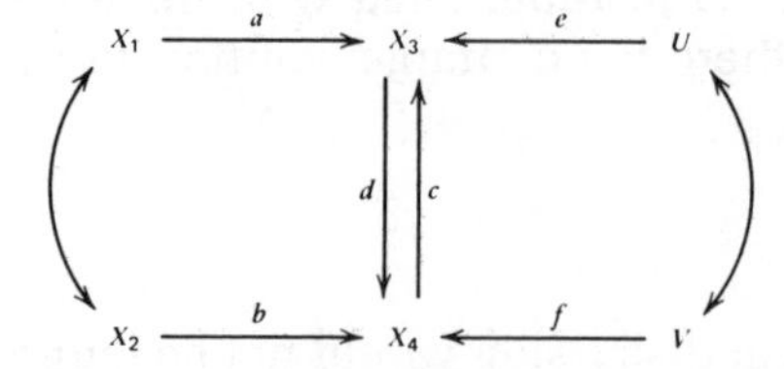

Figure 6.5 Two-stage least-squares example.

Solve for a and c to obtain

$$a = \frac{\rho_{13}\rho_{24} - \rho_{23}\rho_{14}}{\rho_{24} - \rho_{12}\rho_{14}}$$

and

$$c = \frac{\rho_{23} - \rho_{12}\rho_{13}}{\rho_{24} - \rho_{12}\rho_{14}}$$

To estimate b and d for the X_4-equation a parallel procedure is followed. Using the first law one obtains

$$\rho_{14} = b\rho_{12} + d\rho_{13}$$

and

$$\rho_{24} = b + d\rho_{23}$$

The solutions for b and d are then

$$b = \frac{\rho_{24}\rho_{13} - \rho_{14}\rho_{23}}{\rho_{13} - \rho_{12}\rho_{23}}$$

and

$$d = \frac{\rho_{14} - \rho_{24}\rho_{12}}{\rho_{13} - \rho_{12}\rho_{23}}$$

To find e first note that

$$eU = X_3 - aX_1 - cX_4$$

Taking the variance of eU yields

$$e^2 = 1 + a^2 + c^2 - 2a\rho_{13} - 2c\rho_{34} + 2ac\rho_{14}$$

To solve for e simply take the square root. By a similar set of steps it can be shown that

$$f^2 = 1 + b^2 + d^2 - 2b\rho_{24} - 2d\rho_{34} + 2bd\rho_{23}$$

To find ρ_{UV} simply take the covariance of eU with fV to obtain

$$\rho_{34}(1 + cd) + bc\rho_{24} + ab\rho_{12} + ad\rho_{13} - c - d - a\rho_{14} - b\rho_{23}$$

To solve for ρ_{UV} simply divide by ef.

Ordinarily two-stage least squares is computationally more efficient than the preceding procedure. Also, as was noted in the previous chapter, overidentified models present no difficulty for two-stage least square estimation.

To apply two-stage least squares to an equation for X_i with a set of m exogenous variables and k variables which are directly or indirectly caused by X_i, at least k of the exogenous variables must

1. Not directly cause X_i.
2. Not be correlated with the disturbance of X_i.

These k exogenous variables are called the set of instruments. To avoid empirical underidentification, the regression of each of the k endogenous variables on the instruments must first yield moderate to high multiple correlations and the predicted variables of the second stage must not be highly correlated among each other. The reader should refer to Chapter 5 for a more detailed discussion of these conditions.

To apply two-stage least squares, one exogenous variable must be omitted from the equation of each variable in a feedback loop. This variable can be called the *instrument*. However, the same variable cannot be omitted from both equations. To understand this, examine the path diagram in Figure 6.6 which is taken from Duncan (1975, p. 84). It would seem that X_2 could be used as an instrument for both the X_3 and X_4 equation. However, it cannot be since the model in Figure 6.6 is underidentified. Although there are six free parameters and six correlations, there is a hidden pair of overidentifying restrictions. To find them use the modified tracing rule for feedback models:

$$\rho_{13} = \frac{a + bc}{1 - cd}$$

$$\rho_{14} = \frac{b + ad}{1 - cd}$$

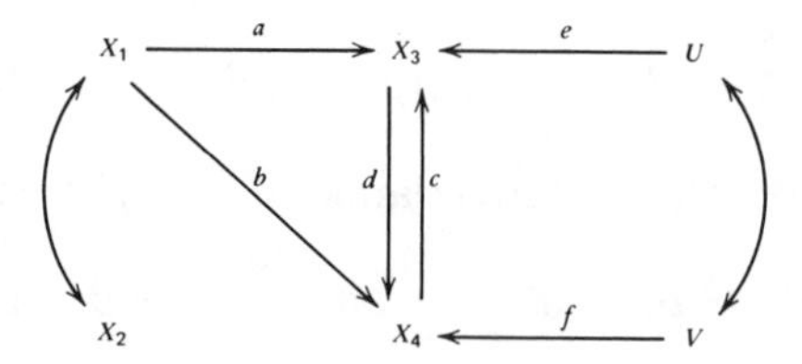

Figure 6.6 Duncan example.

$$\rho_{23} = (\rho_{12})\frac{a + bc}{1 - cd}$$

$$\rho_{24} = (\rho_{12})\frac{b + ad}{1 - cd}$$

The following overidentifying restrictions then hold:

$$\rho_{23} - \rho_{12}\rho_{13} = 0$$

and

$$\rho_{24} - \rho_{12}\rho_{14} = 0$$

These overidentifying restrictions simply state that X_2 shares no variance with X_3 or X_4 except through X_1. Given these overidentifying restrictions, none of the parameters of the model can be estimated. Duncan has noted that the "estimate" of c is

$$\frac{\rho_{23} - \rho_{12}\rho_{13}}{\rho_{24} - \rho_{12}\rho_{14}}$$

and of d is

$$\frac{\rho_{24} - \rho_{12}\rho_{14}}{\rho_{23} - \rho_{12}\rho_{13}}$$

He then points out the curious fact these two estimates are reciprocals of each other. However, neither of these two values are mathematical estimates. Note that, given the overidentifying restriction, the denominators of each expression are zero which makes the formulas mathematical nonsense. Thus neither of the "estimates" are really estimates. The "estimates" of a and b are

$$a = \frac{\rho_{13}\rho_{24} - \rho_{23}\rho_{14}}{\rho_{24} - \rho_{12}\rho_{14}}$$

and

$$b = \frac{\rho_{23}\rho_{14} - \rho_{13}\rho_{24}}{\rho_{23} - \rho_{12}\rho_{13}}$$

Again note that both of the denominators are zero.

A NONHIERARCHICAL MODEL FOR HALO EFFECT

Instead of an empirical example of nonhierarchical models, a conceptual example is presented in this section. Within the topic of person perception in social psychology, it is well known that the ratings of traits and behaviors tend to be biased by the halo effect (Berman & Kenny, 1976). Observers who believe that two traits go together will in fact see the traits as going together whether they in fact do or not.

There are many possible types of models that one could posit for halo. What follows is one such model. Imagine that two traits, friendly and intelligent, have been rated by two raters for a number of ratees. Designate F as the true friendliness of the ratees and I as the true intelligence. Then let F_1 and I_1 be the ratings of the two traits by rater 1 and F_2 and I_2 by rater 2. In Figure 6.7 these six variables are theoretically related in a path diagram. It is first assumed that the ratings of F are indeed caused by F and similarly I causes I_1 and I_2. What brings about halo is that F_1 and I_1 are assumed to be involved in a feedback relationship. So also are F_2 and I_2. Thus the ratings of the two traits mutually influence each other. To simplify the model, F and I are uncorrelated and various equalities of parameters have been assumed in Figure 6.7.

Various authors (Campbell & Fiske, 1959; Stanley, 1961) have stated or implied that the different rater–different trait correlation is free from halo. Thus if the two traits are uncorrelated ($\rho_{FI} = 0$ as in Figure 6.7) then the different rater–different trait correlations should be zero, $\rho_{F_1I_2} = \rho_{F_2I_1} = 0$. This might be true for some models of halo but it is not true for the model in Figure 6.7. Note that both $\rho_{F_1I_2}$ and $\rho_{F_2I_1}$ equal

$$\frac{a^2c + b^2d}{(1 - cd)}$$

Thus, to the extent that both the ratings mutually influence each other (c and d are nonzero) and the raters are to some degree accurate (a and

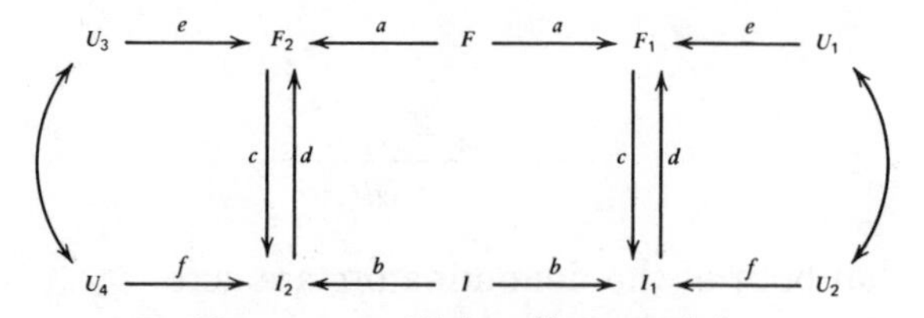

Figure 6.7 Halo effect model.

b are nonzero), the different rater–different trait correlation is nonzero. Thus halo can bias not only the correlations of traits computed within a rater, but also halo can bias correlations of ratings computed from two different raters.

To understand the source of this bias, consider the reduced form equations for F_1 and I_1:

$$F_1 = \frac{a}{(1-cd)}F + \frac{bd}{(1-cd)}I + \frac{df}{(1-cd)}U_2 + \frac{e}{(1-cd)}U_1$$

$$I_1 = \frac{ac}{(1-cd)}F + \frac{b}{(1-cd)}I + \frac{f}{(1-cd)}U_2 + \frac{ec}{(1-cd)}U_1$$

Note that both F and I are in the reduced form equations of both F_1 and I_1. The same holds for F_2 and I_2. The different rater–different trait correlations are nonzero since F and I are in the reduced form equations of the variables.

CONCLUSIONS

My own suspicion is that the strong assumption of equilibrium is sufficiently implausible to make nonhierarchical models generally impractical for the applied social scientists. Added to this the difficulty of finding an instrument, my own view is that feedback models should be applied rarely. (There are notable applications such as Duncan, Haller, and Portes (1971).) Thus, the treatment of nonhierarchical models here has been somewhat sketchy. The reader is referred to excellent introductions of the topic in both Duncan (1975) and Heise (1975).

The next chapter concerns itself with the use of confirmatory factor analysis to estimate structural coefficients.

7

Single Unmeasured Exogenous Variables

Our attention has so far almost exclusively been devoted to measured exogenous variables. In this chapter unmeasured exogenous variables are considered and a factor analysis vocabulary is used as well as a regression analysis vocabulary. Many of the important variables in causal models cannot be directly observed but can be measured only with error. Measured variables are often imperfect glimpses of an underlying reality, and unmeasured variables must be added to the structural model. There are four sections to this chapter. The first section introduces the factor analysis model and discusses the problem of identification. The second section takes up the question of estimation, and the third considers specification error. The final section discusses the measurement of factors.

FACTOR MODELS

Identification

Our initial concern is with the *factor model* and not with the estimation of parameters of the model or, as it is usually called, *factor analysis*. In a factor model the measured variables are assumed to be a linear function of a set of unmeasured variables. These unmeasured, latent variables are called *factors*. Traditionally, one distinguishes three different types of factors: common, specific, and error. A *common factor* is an unmeasured variable that appears in the equation of two or more measured variables. A *specific factor* appears in a single equation and represents true variance that is not common to the other variables. An *error factor* is simply the errors of measurement of a single variable. Procedurally it is difficult to distinguish between specific and error factors, and so they are usually lumped together to form the *unique factor*.

There are two different types of parameters in factor models: *factor loadings* and *factor correlations*. A factor loading is the regression coefficient of a measured variable on a factor. If both the factor and measured variable are standardized, the loading is a beta weight. The factor loadings are multiplied by the factors and summed to equal the measured variable. The variable is said to *load* on a factor. In special cases the factor loading can equal the correlation of the variable with the factor.

The factors themselves may be intercorrelated. The correlations are called factor correlations. Traditionally, the unique factors are uncorrelated with the common factors and with each other. As is demonstrated in this chapter, the unique factors may be correlated. If the common factors are uncorrelated with each other, the solution is said to be orthogonal, and if correlated, it is said to be oblique.

The variance of each measured variable can be partitioned into three components: common, specific, and error. The common variance, called the *communality*, is the squared multiple correlation of the common factors with the measured variables, and its usual symbol is h^2. If the common factors are orthogonal (that is, uncorrelated), then the communality simply equals the sum of squared common factor loadings. The communality plus the specific variance equals the reliability of a variable and the remaining variance is due to error. The specific and error variance may be lumped together to form the unique variance.

In this chapter, we see that with causal models the usual bugaboos of both choosing communalities and choosing a rotation scheme are avoided. Additionally, it is shown that the common factors take the role of exogenous variables, the unique factors of disturbances, the factor loadings of path coefficients, and the factor correlations of correlations between exogenous variables.

One Factor, Two Indicators

Single-factor models presume that the measured variables are all indicators of a latent construct. Using the vocabulary of test theory introduced in Chapter 5, the measures are a set of congeneric tests. Measures that correlate more highly with the latent construct have proportionately less unique variance and can be used to name the construct. In the limiting case a correlation of one implies the measure and the construct are one in the same.

In Figure 7.1 there is a causal model in which X_1 and X_2 are caused by F; U causes X_1, and V causes X_2. If X_1 and X_2 are measured, all the measures are endogenous variables, all the causes are unmeasured, and the relationship between X_1 and X_2 is spurious. What is the status of identification of the model in Figure 7.1? There is only a single correlation, ρ_{12}, but there are two free causal parameters, a and b. As with models from the previous chapters, the paths from the disturbances are constrained since $c = (1 - a^2)^{1/2}$ and $d = (1 - b^2)^{1/2}$. Since there are more parameters than correlations, the model is underidentified. The parameters of the model cannot be estimated, but both a^2 and b^2 must be at least $\rho_{12}{}^2$. To achieve identification it might be assumed that $a = b$. It would then follow that $a = b = \rho_{12}{}^{1/2}$. However, if the value of ρ_{12} were negative, it would contradict any assumption that $a = b$ since a parameter would have to be imaginary, the square root of minus one.

The units of measurement of the factor are arbitrary. The approach taken here, which is consistent with the factor analysis tradition, is to standardize the factor. An alternative approach is to use the units of one of the measures. In this case a loading of one measure is set to one and the variance of the factor is free.

One Factor, Three Indicators

The status of identification changes if there are three measured variables that are caused by the latent variable, F. The measures X_1, X_2, and X_3 are sometimes called the *indicators* of the latent construct. Also the correlation of an indicator with a factor is called an *epistemic correlation* (Costner, 1969) and is the factor loading. As shown in Chapter 3, given

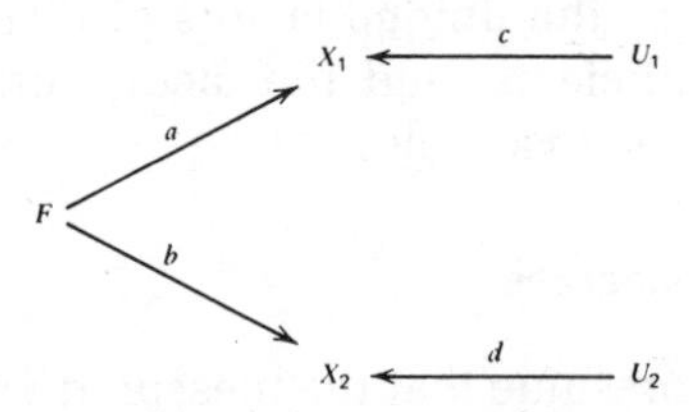

Figure 7.1 Single-factor model, two indicators.

$$X_1 = aF + dU_1$$
$$X_2 = bF + eU_2$$
$$X_3 = cF + fU_3$$

and

$$\rho_{FU_i} = 0 \ (i = 1,2,3)$$
$$\rho_{U_iU_j} = 0 \ (i \neq j)$$

it follows from the tracing rule that $\rho_{12} = ab$, $\rho_{13} = ac$, and $\rho_{23} = bc$. Since there are three parameters and three correlations, the model is just-identified. The estimates are

$$a^2 = \frac{\rho_{12}\rho_{13}}{\rho_{23}}$$

$$b^2 = \frac{\rho_{12}\rho_{23}}{\rho_{13}}$$

$$c^2 = \frac{\rho_{13}\rho_{23}}{\rho_{12}}$$

The disturbance paths are simply

$$d = (1 - a^2)^{1/2}$$
$$e = (1 - b^2)^{1/2}$$
$$f = (1 - c^2)^{1/2}$$

The direction of the scale of the factor is arbitrary. One can then take either the positive or negative root of a^2 or any of the other parameters. As a rule one normally tries to make as many loadings as possible positive. Once the sign of one parameter has been set, the sign of the others have been determined. For instance, if a is taken to be positive and if ρ_{12} is negative and if ρ_{13} is positive, b must be negative and c positive. However, if a is taken to be negative, then b is now positive and c is negative. In either case, ρ_{23} must be negative, and if it is not, some specification error is indicated. In general, it must hold that $\rho_{12}\rho_{13}\rho_{23} > 0$ for there to be no contradiction about the direction of the scale of the unmeasured variable. With three variables for there to be no contradiction all or only one of the correlations should be positive. If $\rho_{12}\rho_{13}\rho_{23} < 0$, one of the factor loadings must be imaginary.

Thus with three indicators the single factor model is identified. However, if any of the loadings (a, b, or c) are zero, or near zero, the model is empirically underidentified. For instance, let a equal zero and note that the estimates for b^2 and c^2 are undefined since $\rho_{13} = \rho_{12} = 0$. Thus to identify empirically the loadings on a single factor, the loadings must be of moderate size.

It is also mathematically possible that the value obtained for a^2 or any loading to be larger than one. For a^2 to be larger than one, it follows that $|\rho_{12}\rho_{13}| > |\rho_{23}|$. Although a value larger than one is mathematically possible, in the population it indicates a specification error. Since a is the correlation of F with X_1, it cannot be greater than one. Although correlations *computed* from data cannot be larger than one, correlations constructed from data can be out of bounds. If a^2 is larger than one, then the solution for d is imaginary. Estimates of parameters larger than one in factor analysis are examples of the Heywood case (Harmon, 1967, p. 117).

Using the sample data, it is possible to obtain an estimate of a loading larger than one or imaginary, even if the model is correctly specified. A method then is needed to test whether the anomalous estimate is due either to sampling error or to specification error. With a little bit of algebraic manipulation it can be shown that both imaginary and loadings larger than one occur only in the sample if $r_{12.3}$, $r_{13.2}$, or $r_{23.1}$ is negative. Thus a test that the anomaly is plausibly explained by sampling error is that a negative partial correlation does not significantly differ from zero. If it is significantly negative, then specification error is indicated.

One Factor, Four Indicators

In Figure 7.2 there is one common factor, F, and four endogenous variables, X_1 through X_4. The disturbances, U_1 through U_4, are assumed

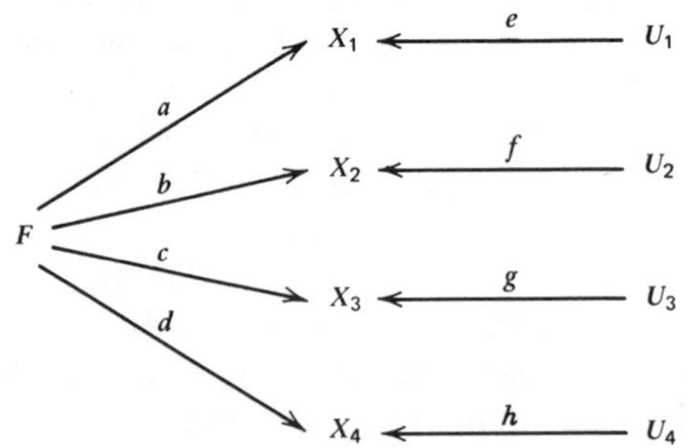

Figure 7.2 Single-factor model, four indicators.

to be uncorrelated with each other and with F. The model is overidentified in some way since there are four free parameters and six correlations. Using the tracing rule, the six correlations are

$$\rho_{12} = ab$$

$$\rho_{13} = ac$$

$$\rho_{14} = ad$$

$$\rho_{23} = bc$$

$$\rho_{24} = bd$$

$$\rho_{34} = cd$$

One way to solve for the parameters is to view this four-variable model as a set of four different three-variable models; that is, separately analyze variables X_1, X_2, and X_3, then analyze X_1, X_2, X_4, then, X_1, X_3, X_4, and finally, X_2, X_3, X_4. Then separately solve for the parameters from each of these three variable cases as was done in the previous section. Working through this strategy yields the following solutions for each parameter:

$$a^2 = \frac{\rho_{12}\rho_{13}}{\rho_{23}} = \frac{\rho_{12}\rho_{14}}{\rho_{24}} = \frac{\rho_{13}\rho_{14}}{\rho_{34}}$$

$$b^2 = \frac{\rho_{12}\rho_{23}}{\rho_{13}} = \frac{\rho_{12}\rho_{24}}{\rho_{14}} = \frac{\rho_{23}\rho_{24}}{\rho_{34}}$$

$$c^2 = \frac{\rho_{13}\rho_{23}}{\rho_{12}} = \frac{\rho_{13}\rho_{34}}{\rho_{14}} = \frac{\rho_{23}\rho_{34}}{\rho_{24}}$$

$$d^2 = \frac{\rho_{14}\rho_{24}}{\rho_{12}} = \frac{\rho_{14}\rho_{34}}{\rho_{13}} = \frac{\rho_{24}\rho_{34}}{\rho_{23}}$$

There is an interesting pattern in the subscripts of the preceding formulas. Although it makes no sense algebraically, cancel the subscripts of the correlations. For instance, for the first expression for a^2, the 2 and 3 cancel leaving two 1s. In every case two subscripts of the endogenous variable remain. Using "subscript algebra," all that remains is the subscript squared which provides a perfect analog to the parameter estimate which is also squared. I do not recommend subscript algebra as a replacement for careful analysis, but it can serve as a useful mnemonic and check.

The disturbance paths can be simply estimated by

$$e = (1 - a^2)^{1/2}$$

$$f = (1 - b^2)^{1/2}$$

$$g = (1 - c^2)^{1/2}$$

$$h = (1 - d^2)^{1/2}$$

Since there are multiple estimates for each parameter, overidentifying restrictions can be solved for. All the preceding equalities of the parameter solutions imply only the three following overidentifying restrictions:

$$\rho_{12}\rho_{34} - \rho_{13}\rho_{24} = 0 \qquad [7.1]$$

$$\rho_{12}\rho_{34} - \rho_{14}\rho_{23} = 0 \qquad [7.2]$$

$$\rho_{13}\rho_{24} - \rho_{14}\rho_{23} = 0 \qquad [7.3]$$

since $\rho_{12}\rho_{34} = \rho_{13}\rho_{24} = \rho_{14}\rho_{23} = abcd$. It can be quickly seen that if any of the two preceding restrictions holds, the third immediately follows. There are then only two independent overidentifying restrictions which was already known since there are four free parameters and six correlations. This excess of correlations yields two independent overidentifying restrictions, that is, two constraints on the correlation matrix.

These overidentifying restrictions are called *vanishing tetrads*. They are called *vanishing tetrads* since four variables are involved and the difference between pairs of products equals zero; that is, they vanish. Note that the tetrads satisfy subscript algebra. For instance, rearranging Equation 7.1, the following holds

$$\frac{\rho_{12}\rho_{34}}{\rho_{13}\rho_{24}} = 1$$

and the subscripts all cancel to yield one.

One Factor, N Indicators

In the general case, given n measures and a single factor model, there are $(n - 1)(n - 2)/2$ estimates of each parameter. Since with every four variables there are three tetrads, there are a total of $3\binom{n}{4}$ tetrads or

$$\frac{n(n - 1)(n - 2)(n - 3)}{8}$$

The total number of independent tetrads is the number of correlations, $n(n-1)/2$, minus the number of free parameters, n, which yields $n(n-3)/2$ restrictions on the correlations.

Vanishing tetrads are very common overidentifying restrictions in models with unobserved variables. More often than not if there is an overidentifying restriction, it is a vanishing tetrad.

Test of a Vanishing Tetrad

Deferred until this chapter is the significance test of a vanishing tetrad. Duncan (1972) resurrected a test for the vanishing tetrad which was originally suggested by Spearman and Holzinger (1924). The test is not well known and yields a different value depending on the order of the products of the correlations. I have suggested a test for the vanishing tetrad (Kenny, 1974) that can be made by using a canned computer program available at most installations: canonical correlation. The use of canonical correlation analysis for models with unmeasured variables has been suggested by other authors (e.g., Hauser & Goldberger, 1971), but none had explicitly suggested the canonical correlation as a test for the vanishing tetrad.

For those who may be unfamiliar with canonical correlation, there are two sets of variables, say $X_1, X_2, \ldots, X_p$ and a second set $Y_1, Y_2, \ldots, Y_q$. There are then p variables in set X and q in set Y. The canonical correlation is the correlation between a linear combination of set X with a linear combination of set Y where the weights for both X and Y are set to maximize that correlation. If both p and q are greater than one, a second canonical correlation can be obtained such that the new linear combinations are orthogonal to the prior combinations. The total number of canonical correlations is equal to p or q, whichever is smaller.

Canonical correlation can be used to test a vanishing tetrad of $\rho_{13}\rho_{24} - \rho_{14}\rho_{23} = 0$ (Equation 7.3). Let X_1 and X_2 be set X and X_3 and X_4 be set Y. To determine how the four variables divide into the sets note that the tetrad does not involve ρ_{12} and ρ_{34}. The simple rule is that X_1 and X_2 form one set (ρ_{12}) and X_3 and X_4 (ρ_{34}) the second set. For the tetrad of Equation 7.1 the missing correlations are ρ_{14} and ρ_{23}. Thus X_1 and X_4 form one set and X_2 and X_3 the second set.

It can be shown that if $r_{13}r_{24} - r_{14}r_{23} = 0$, then the second canonical correlation will be zero. Thus the null hypothesis, Equation 7.3, can be tested by the hypothesis that the second canonical correlation is zero. In the same way the remaining tetrads can be tested. A significance test for this special case of $p = q = 2$ is given by Bartlett's χ^2 approximation (Tatsuoka, 1971, p. 189):

$$\chi^2 = -(N - 3.5) \log_e(1 - r_2^2) \qquad [7.4]$$

where N is the sample size and r_2 is the estimated second canonical correlation and the χ^2 has one degree of freedom.

In general the computation of canonical correlations is rather involved, but a handy solution is relatively simple when both p and q are two. Solve the following quadratic equation for ρ^2

$$a\rho^4 + b\rho^2 + c = 0 \qquad [7.5]$$

where

$$a = (1 - r_{12}^2)(1 - r_{34}^2)$$

$$b = (r_{14} - r_{12}r_{24})(r_{13}r_{34} - r_{14}) + (r_{23} - r_{13}r_{12})(r_{24}r_{34} - r_{23}) + (r_{13} - r_{12}r_{23})(r_{14}r_{34} - r_{13}) + (r_{24} - r_{14}r_{12})(r_{23}r_{34} - r_{24})$$

$$c = (r_{13}r_{24} - r_{14}r_{23})^2$$

Since Equation 7.5 is quadratic, there are two solutions for ρ^2. The second canonical correlation squared is set equal to the smaller solution of Equation 7.5. If c equals zero, then the smaller solution of ρ^2 of Equation 7.5 is zero. Note that c exactly equals the square of a tetrad.

As an illustration, consider the correlations from Duncan (1972, p. 53): $r_{12} = .6508$, $r_{13} = .7304$, $r_{14} = .7548$, $r_{23} = .7100$, $r_{24} = .6999$, and $r_{34} = .6642$, where the sample size is 730. A test of the hypothesis that $\rho_{13}\rho_{24} - \rho_{14}\rho_{23} = 0$ yields a second canonical correlation of .050 and $\chi^2(1) = 1.80$, which is not significant at the .05 level. The solution of Equation 7.5 yields $a = .322$, $b = -.247$, and $c = .0006$.

ESTIMATION

The preceding sections present only the path analytic estimates of the path from the factor to the measure. As the reader is no doubt aware, there are a myriad of possible estimates of factor loadings from different methods of factor analysis. Before discussing the mechanics of estimating path coefficients, consider a data set to be used as an illustration. The data are taken from unpublished data of B. Malt. One hundred eighteen terminal cancer patients were asked to state the strength of their concerns about a number of issues. The four concerns are concerns about friends (X_1), concerns about existence (X_2), concerns about family (X_3), and concerns about self (X_4). The correlations between these four variables are contained in Table 7.1. A path model for the four variables, contained in Figure 7.2, has the four specific concerns caused by some unmeasured general concern.

A single factor model adequately fits the data since the three tetrads appear to vanish

$$(.242)(.719) - (.551)(.416) = -.055$$

$$(.242)(.719) - (.577)(.311) = -.005$$

$$(.551)(.416) - (.577)(.311) = .050$$

The fact that the three preceding tetrads do not perfectly vanish can be explained by sampling error since the χ^2 for the tetrads are 1.00, .01, and 1.42, respectively. None of the χ^2 values even approaches statistical significance. Remember if the χ^2 is not significant, the tetrad vanishes and a single-factor model is indicated.

Since a single-factor model fits the data, the next question is, what are the path coefficients from the factor to each measure? Earlier it was stated that these path coefficients are factor loadings. A comparison of four different methods of estimating the loadings follows:

1. Pooled path analytic.
2. Principal components.
3. Principal factors with iterated communalities.
4. Maximum likelihood.

Earlier it was shown that there are three different formulas for path analytic solutions when there are four measured variables. If one sub-

Table 7.1. Four Concerns of Terminal Cancer Patients: Correlations below Diagonal and Residuals from the Maximum Likelihood Solution above the Diagonal[a]

	X_1	X_2	X_3	X_4
Concern about friends (X_1)	1.000	−.040	.027	−.008
Existential concern (X_2)	.242	1.000	−.037	.028
Concern about family (X_3)	.551	.311	1.000	−.002
Concern about self (X_4)	.577	.416	.719	1.000

[a]N = 118.

stitutes in the sample estimates of the correlations, the estimates of the loadings for the correlations in Table 7.1 are

a: .655, .579, .665

b: .370, .418, .424

c: .841, .829, .733

d: .996, .868, .981

These different estimates of the same parameter are fairly homogeneous the largest difference being only .128. This is not surprising since the overidentifying restrictions were met. When the tetrads vanish, the different estimates are exactly the same. Now the problem arises of how to combine these parameter estimates. One solution might be a simple arithmetic average. This is a flawed strategy since it fails to weight the estimates by the size of the correlations. An alternative strategy is recommended by Harmon (1967): Examining the estimates of a^2 note that they all equal the product of two correlations divided by a third. Why not simply pool the numerators and the denominators separately? Then to estimate a^2 divide the pooled numerator by the pooled denominator to obtain

$$\frac{r_{12}r_{13} + r_{12}r_{14} + r_{13}r_{14}}{r_{23} + r_{24} + r_{34}}$$

The numerator involves all possible pairs of correlations with variable X_1, and the denominator is the sum of all correlations not involving X_1. This procedure assumes that all the correlations are positive in sign. If some of the correlations are negative, it is advisable to sum absolute values. In a similar fashion we can pool the estimates for the other path coefficients. These pooled path analytic estimates for the correlations in Table 7.2 are in Table 7.2. They fall between the three path analytic values. It is interesting to note that the loading for concerns about self, or X_4, is almost equal to one. Thus, the unmeasured underlying concern is very close to a concern about the self.

The second solution discussed is perhaps the best known: the method of principal components. The nonmathematical logic of this solution is as follows: Let us create a linear composite of the measured variables, choosing the weights of this composite so that we maximize the sum of squared correlations of the variables with the composite. Since a squared correlation can be thought of as variance explained, we choose the composite to maximize its ability to explain the variance of

Table 7.2. Factor Loadings for the Correlations in Table 7.1

	Solution			
Variable	Pooled Path Analytic	Principal Components	Iterated Communalities	Maximum Likelihood
X_1	.639	.774	.650	.652
X_2	.407	.568	.415	.432
X_3	.800	.857	.805	.804
X_4	.930	.893	.906	.897

the measured variables. Mathematically this sort of problem turns out to be an *eigenvector* problem, which is computationally a bit messy but tractable. The resultant correlation of the composite with each measure is the factor loading. There is clearly something backward about a principal component solution, at least when one is doing causal modeling. With principal components the factor is directly measurable since it is defined as a weighted sum of the measured variables. In causal modeling, each measured variable or any finite sum of them is only an imperfect indicator of the factor.

The factor loadings from principal components are in Table 7.2. Although their relative ranking compares favorably to the pooled path analytic solution, they are hardly the same solution. The most striking fact about the principal component solution is that in three of the four cases the factor loadings are larger than the pooled path analytic solution. If we use the principal components solution to estimate the correlations between the measured variables, we overestimate all six of the correlations. *The components solution does not attempt to fit the correlations, but rather it attempts to maximize the sum of squared factor loadings.* Since this is the goal, a principal components solution does not provide a very good estimate of path coefficients.

To obtain the true factor loadings the communalities should be placed in the main diagonal of the correlation matrix before the eigenvectors are computed. But since the factor loadings are needed to compute the communalities, there is a chicken–egg problem. We need the communalities to compute the factor loadings and we need the loadings to compute the communalities. One way around this problem is to find the communalities by *iteration*. Start with some communality estimates and then solve for the factor loadings by the method of principal factors. (A *principal factors* solution attempts to maximize the explanation of only the common variance, not all the variance as does principal components.) Taking these factor loadings we solve for communalities, and with these new communalities we solve for the factor loadings. We continue until the factor loadings stabilize. To hasten the convergence, the initial communalities are the squared multiple correlations of each variable with the other variables taken as the predictors. The principal factors solution with iterated communalities is an option in SPSS (Nie, Hull, Jenkins, Steinbrenner, & Bent, 1975).

The iterated communality solution for the factor loadings is contained in Table 7.2. The values more closely resemble the pooled path analytic solution than the principal components solution. The solution does result in somewhat less variable loadings than the path analytic solution.

The final solution discussed is *maximum likelihood* factor analysis. Of the four solutions only maximum likelihood provides a statistical estimate of factor loadings or path coefficients and only maximum likelihood factor analysis provides significance tests of goodness of fit of the model. A *likelihood* is the analog of probability for continuous distributions. One can compute the likelihood of a set of estimates of factor loadings given the sample correlation matrix and the assumption of multivariate normality. The method of maximum likelihood chooses the most likely estimates. Like the previous method, it is also an iterative solution. Although the mathematics of maximum likelihood factor analysis may remain a "black box" for most of us, suffice it to say the computations have proved to be tractable only by computer. The pioneering work was done by Lawley and Maxwell (1963), and the major computational breakthrough was made by Jöreskog. There is little doubt that maximum likelihood factor analysis will revolutionize factor analysis since it puts factor analysis estimation on a firm statistical ground.

Unlike most traditional methods of factor analysis, maximum likelihood factor analysis can be employed in either a confirmatory or an exploratory manner. Most methods of factor analysis can only be employed in exploratory fashion; that is, little or no structure can be imposed on the solution. In the case of a model with multiple factors the solution is not unique and can be rotated to an infinite number of alternative solutions. If the maximum likelihood solution is combined with a confirmatory approach, the solution is unique and no rotation of the solution is possible.

If a confirmatory approach is chosen, the model must be identified. The researcher must impose enough constraints to make the parameter estimates unique. The two common types of constraints are fixing the parameters to a given value and forcing two parameter estimates to be equal. With maximum likelihood factor analysis one can fix factor loadings, factor variances, factor correlations, factor covariances, or communalities to a given value. Usually the fixed value is zero or one. For instance, a loading or a factor correlation is set to zero or a loading or factor variance is set to one. Equality constraints are also possible: Two loadings or factor correlations may be set equal to each other. In this text maximum likelihood factor analysis is always employed in a confirmatory manner. Thus maximum likelihood factor analysis is at times referred to as confirmatory factor analysis.

In Table 7.2 is the maximum likelihood solution. It is closest to the iterated communality solution. Its estimated factor loadings are less variable than any of the other solutions. Maximum likelihood judges the .242 correlation between X_1 and X_2 to be too small, which raises the estimate of the loadings for those two variables compared to the pooled path analytic and iterated communality solutions. In general, the solution with iterated communalities provides a solution very close to the maximum likelihood solution.

In confirmatory factor analysis for overidentified models one can also test the fit of the model to the observed correlations. The test is a χ^2 goodness of fit test with as many degrees of freedom as the number of independent overidentifying restrictions. If the χ^2 is not significant then the model fits the data. Thus one is testing the null hypothesis. The χ^2 test is an approximation that is valid only for large samples. For the correlations in Table 7.1 the χ^2 is 1.94 and is not significant, which indicates that a single-factor model satisfactorily fits the data. One can view the χ^2 as the combined test of the vanishing tetrads. In Table 7.1 above the main diagonal, we have the residuals from the maximum likelihood solution that is, the observed correlations minus the predicted correlations. For a single-factor model the predicted correlations between two variables is simply the product of the standardized factor loadings.

The computer program used to obtain the maximum likelihood factor loadings is called Analysis of Covariance Structures or ACOVS (Jöreskog, Gruvaeus, & van Thillo, 1970). A second Jöreskog program called COFAMM can also test for single factoredness as well as a BMDP program.

If one intends to compare loadings across populations or over time, one ordinarily should estimate the *unstandardized* loadings. One would fix the loadings of one of the measures to some arbitrary nonzero value, usually one, and the remaining loadings and factor variance would then be free. Since an unstandardized solution is desired, a variance-covariance should be the input to the factor analysis. ACOVS or COFAMM accepts covariance matrices. If one has obtained the standardized solution, one can destandardize in the following way. For the above example assume for illustrative purposes $V(X_1) = 2$, $V(X_2) = 3$, $V(X_3) = 2$, and $V(X_4) = .5$. If we fix the unstandardized loading of X_1 on F as one, it follows from the standardized solution that

$$r_{1F} = \frac{C(X_1,F)}{(V(X_1)\,V(F))^{1/2}} = .652$$

and from the unstandardized solution that

$$\frac{C(X_1,F)}{V(F)} = 1.000$$

Solving through for $V(F)$ yields

$$V(F) = 2(.652^2) = .850$$

the loadings for the other variables can now be destandardized to yield

$$r_{2F}\left(\frac{V(X_2)}{V(F)}\right)^{1/2} = .811$$

$$r_{3F}\left(\frac{V(X_3)}{V(F)}\right)^{1/2} = 1.233$$

$$r_{4F}\left(\frac{V(X_4)}{V(F)}\right)^{1/2} = .688$$

Standardized loadings are often more interpretable, but the unstandardized loadings should, in principle, replicate better across populations and occasions.

A second illustration is taken from the doctoral dissertation of Hamilton (1975). One hundred ninety persons viewed an audio-visual presentation of a military court martial. The case, adapted from a real case (the names were changed to protect the guilty), concerned a soldier charged with the crime of killing four Korean civilians. After viewing the case, 190 observers were asked to state a verdict (X_1), choose a sentence (X_2), rate the soldier's innocence (X_3), and rate his responsibility (X_4). The intercorrelation matrix of the four variables is in Table 7.3. As might be expected, innocence correlates negatively with the three other variables. If possible, it is useful to *reflect* or reverse the scale of certain variables to obtain all positive correlations.

Table 7.3. Judgments of Persons in Mock Trial: Correlations below the Main Diagonal and Residuals above the Diagonal[a,b]

	X_1	X_2	X_3	X_4
Verdict (X_1)	1.000	−.001	−.008	−.012
Sentence (X_2)	.412	1.000	.042	.055
Responsibility (X_3)	.629	.403	1.000	.001
Innocence (X_4)	−.585	−.270	−.500	1.000

[a]N = 190.
[b]Data taken from Hamilton (1975).

This is not done here to show that negative factor loadings in no way interfere with the analysis.

In Table 7.4 are the four different sets of estimates of the path coefficients for the single-factor model. Again only the principal components solution markedly differs from the three other solutions. Again it tends to overestimate the loadings. The other three sets of estimates are very similar, and again the iterated solution is closer to the maximum likelihood solution than the pooled path analytic solution. Also the maximum likelihood solution is the least variable. In Table 7.3 above the main diagonal we have the residuals. Except for the sentence variable (X_3), they are small.

The test of tetrads 7.1, 7.2, and 7.3 yields χ^2 values of .99, .64, and 2.58, respectively, none of which are statistically significant. The goodness of fit χ^2 for the maximum likelihood solution is 2.62 which also is not significant.

It should be repeated that in this chapter the specification has been made to give the factor unit variance. Such a specification is consistent with traditional factor analysis models, but it is inconsistent with the common practice in the structural modeling literature. There the usual

Table 7.4. Factor Loadings for the Correlations in Table 7.3

	Solution			
Variable	Pooled Path Analytic	Principal Components	Iterated Communalities	Maximum Likelihood
X_1	.860	.865	.854	.854
X_2	.475	.631	.481	.484
X_3	.779	.830	.762	.746
X_4	−.637	−.767	−.656	−.672

practice is to designate one of the indicator variables as a *marker*. Its loading is then fixed to one and the remaining loadings are then free as well as the variance of the factor. Such a strategy is much more valid if the researcher either intends to compare loadings over time, compare the loading of different populations, or desires to fix error variances equal, that is, assumes the measures are tau equivalent.

Convergent and Discriminant Validation

It is unfortunate that researchers rarely check for single factoredness. To some degree this is a legacy due to Spearman's two-factor theory of intelligence. Spearman argued that all cognitive measures can be characterized by two factors. One is the g factor of general intelligence, and the second factor is unique. Although called the "two-factor theory of intelligence," a consequence of the theory is that measures of cognitive skills should be single factored. Subsequent research showed that Spearman's theory did not hit the mark, and since then there has been an overabundance of emphasis on multiple-factor solutions.

Tests of single factoredness can demonstrate discriminant validity. Brewer, Campbell, and Crano (1970) argue that, before causal models of the type discussed in Chapter 4 are entertained, it should be demonstrated that the measures are not all indicators of the same construct. Their argument is especially strong for the case in which all the measures were obtained by the same method. There is a disappointing reluctance to follow the Brewer et al. suggestion. It seems that researchers are not willing to test a model that is inconsistent with their hypotheses. One should, however, remember that by its very nature a null hypothesis is inconsistent with the experimental hypothesis.

Consider the study by Schwartz and Tessler (1972) on the relationship between attitudes and behavior. The dependent variable of their study is behavioral intention to donate an organ for transplant. The independent variables are attitude, personal norms, and social norms. Using multiple regression Schwartz and Tessler conclude that personal norms are the major cause of behavioral intention followed by attitude and, finally, social norms. Since all the measures are self-report, it is plausible that all the measures tap the same latent variable. It is not inconceivable that one's stated attitude, behavioral intention, and report of personal and social norms are all indicators of some underlying disposition. One might posit that the four measures are caused by a single factor. In Table 7.5 are the χ^2 tests of goodness of fit of such a single-factor model for six transplant operations. Since four parameters are estimated (the factor loadings of the four variables) and since there

Table 7.5. Schwartz and Tessler Transplant Data[a]

	χ^2
Kidney	
Relative	1.01[b]
Stranger	6.47[c]
Heart	
Relative	.97
Stranger	3.13
Marrow	
Relative	2.49
Stranger	15.86[c]

[a]N = 195.
[b]Degrees of freedom equals two.
[c]$p < .05$.

are a total of six correlations between the four variables, the χ^2 goodness of fit tests has two degrees of freedom. For four of the six transplant operations a single-factor model is consistent with the data; that is, the χ^2 is not significant. In the giving of kidney and marrow to a stranger cases, the single-factor model is inconsistent with the correlations. In all six instances the estimated factor loadings are greatest for behavioral intention variable (.833 to .916) which suggests that the underlying factor reflects behavioral intentions more closely than it does attitude or norms. There is some support for Schwartz and Tessler's use of multiple regression but no overwhelming support. Discriminant validation is not clearly indicated.

SPECIFICATION ERROR

Although all the models that have been elaborated have nicely fit a single-factor model, unfortunately this does not always happen. When the model does not fit the data, some sort of specification error is indicated. (Moreover even if the model fits, a simpler model may fit better.) An additional specification of *correlated disturbances* (that is, correlated unique factors) can be made to salvage a single-factor model. Correlated disturbances imply that there is an additional source of covariance between two measures besides the factor. A researcher can specify as many correlated disturbances as there are overidentifying restrictions, but the researcher must be careful about which are made since certain patterns of correlated disturbances may yield an underidentified model. For instance, with four variables there are two independent overidentifying restrictions, and so two sets of disturbances can be correlated. If the disturbances of X_1 and X_2 and also of X_3 and X_4 are correlated, the model is underidentified. As shown in Chapter 3, an overidentifying restriction holds:

$$\rho_{13}\rho_{24} - \rho_{14}\rho_{23} = 0$$

The overidentifying restriction plus the six free parameters is one more than the six correlations. Thus, if correlated disturbances are added to the model, one must be careful about the status of identification. In general, to identify an n-variable single-factor model with correlated disturbances, there must be first at least three variables whose disturbances are not intercorrelated and, second, for every other variable, its disturbance must be uncorrelated with at least one other variable whose factor loading can be estimated. Some patterns of correlated disturbances are also illogical. For instance, if the disturbance of X_1 is correlated with X_2 and X_3, more often than not one should have the disturbances of X_2 and X_3 be correlated. It makes almost no sense to have them uncorrelated.

Ideally the correlated disturbances can be specified by theory. More often they are unanticipated and may be indicated by large residuals. Of course, adding correlated disturbances after examining residuals makes the data analysis more exploratory than confirmatory. Moreover, examination of the residuals may be misleading (Costner & Schoenberg, 1973).

Recall that the overidentifying restriction of a single-factor model is a vanishing tetrad. If the disturbances of any two variables are allowed to be correlated, then the correlation between those two variables cannot be used in any tetrad. Thus, if the disturbances of X_1 and X_2 are correlated, tetrads 7.1 and 7.2 do not necessarily hold. However, the fact that only tetrad 7.3 holds does not necessarily imply that the disturbances of X_1 and X_2 are necessarily correlated. If the disturbances of X_3 and X_4 are correlated, again only tetrad 7.3 holds. Thus, one cannot necessarily use the tetrads to determine exactly which disturbances are correlated. The data always are consistent with a multitude of structural models.

If correlated disturbances are specified, the model does not necessarily always recover the sample variances. Thus, there are residuals not only for the correlations (covariances) but also for the variances.

A final type of specification error cannot be tested. In Figure 7.3 is a model somewhat different than the model in Figure 7.2. The variable X_1

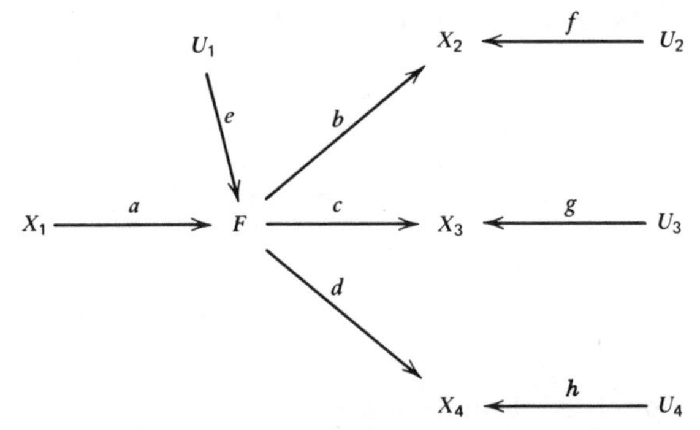

Figure 7.3 One indicator as a cause.

causes F whereas in Figure 7.2 F causes X_1. There is, however, no way to distinguish the two models empirically. They both have the same overidentifying restrictions. One might then reinterpret the correlations in Table 7.1 as showing that concern about one's self causes some latent variable which in turn causes concerns about family, friends, and life.

Hierarchical Model Testing

It is possible to compare models in confirmatory factor analysis to judge whether the fit improves when more parameters are added to the model or, alternatively, to judge whether the fit worsens when parameters are deleted from the model. This procedure of comparing models is similar to model comparison methods in analysis of variance (Appelbaum & Cramer, 1974), multiple regression, and log-linear models for contingency tables (Fienberg, 1977). In Table 7.6 are a series of single-factor models. Near the top of the tree note the entry, *free loadings*. This represents a model in which the loadings for each variable are free to vary. Below that entry is a model with equal loadings. This model is more restrictive than the model above it, since a model with equal loadings is a special case of a model with free loadings. Still more restrictive is a model in which the loadings are all fixed equal to some value, say zero. A model with fixed equal loadings is a special case of a model with free equal loadings.

If there are two models, say A and B, and model A is more restrictive than B, the test of the restrictions added to model A can be obtained by subtracting the χ^2 for model B from the χ^2 for model A. The difference between the χ^2 values is itself a χ^2 with degrees of freedom equal to the

Table 7.6. A Hierarchy of Models

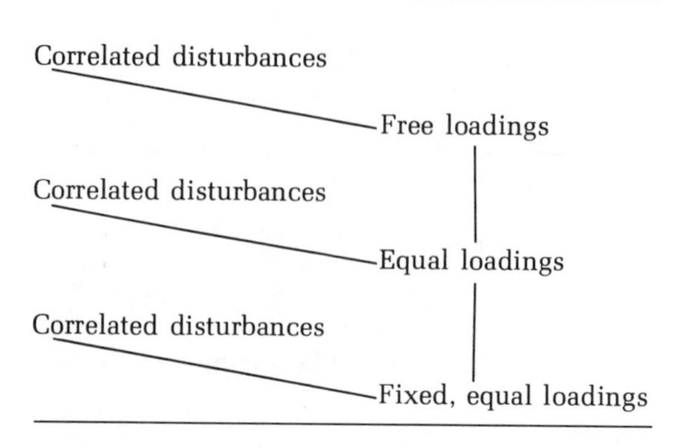

difference in the degrees of freedom of the two χ^2 values. For example, to test the additional constraints of equal loadings over free loadings, one takes the smaller χ^2 and subtracts if from the larger. The degrees of freedom of this χ^2 are the corresponding difference in degrees of freedom. If the difference χ^2 is not significant, then one cannot reject the null hypothesis that the factor loadings are equal. Similarly one can compare a model with equal loadings to a model in which the loadings are all fixed at zero. The comparison of the model with free loadings versus a model with loadings equalling zero simultaneously tests equal loadings and zero loadings.

Each of the three models in Table 7.6 has a branch. For each model a pair of disturbances has been correlated, the same pair for each model. One can then test for the existence of correlated errors in three different ways: with free loadings, with equal loadings, and zero loadings. The resultant χ^2 values will each have one degree of freedom but will not necessarily be equal. The same parameter is being tested but it is added to different models. One cannot then speak of a unique test that a certain parameter equals zero. It can be discussed only in the context of model comparison. This is also true of other statistical methods. For example one can test whether a beta weight is zero in multiple regression only in the context of the other predictor variables. The beta weight may be significant with one set and not with another.

Clearly then it is not sufficient to estimate a single model and be satisfied if it fits. Somewhat simpler models may fit the data better, and possibly even more complex models may indicate important specification errors.

FACTOR MEASUREMENT

Actually the goal of most users of factor analysis is not to *test* a factor model but to *measure* factors. A measured factor for a subject is usually called a *factor score*. Very often a model can be estimated and tested without actually measuring the factors. Unfortunately researchers suffer from the delusion that, unless one can assign "real scores" to subjects, the variable is not real. Just because latent variables are not observable does not mean that they are not real. Much in the same way, atoms are not observable, but they are very real, nonetheless.

One approach to factor measurement is multiple regression. Since the correlation between each of the indicators and the factor is known (the factor loadings in the single-factor case), multiple regression can be used to predict factor scores. Thus, the multiple regression prediction equation is used to estimate factor scores since the true factor scores are not available. For a single-factor model the regression weights are proportional to

$$\frac{b_i}{1 - b_i^2} \qquad [7.6]$$

where b_i is the loading of X_i on the factor. If any loading is near one then Equation 7.6 will be very large. A simple rule is that if a loading is .975 or better, then simply take that variable as the factor score. If the loading is less than .975 then all the weights will be less than 20.

Instead of using multiple regression weights to create the factor scores, one could simply add the variables, weighting them equally. The use of equal weights is called *unit weighting*. Regardless of whether regression or unit weights are used, it is useful to derive the correlation of any composite with the factor since the correlation squared measures the reliability of the factor score. Let a_i stand for the weight for variable X_i; then the composite is defined as $\sum_i a_i X_i$. For a composite of measured *standardized* variables, $\sum_i a_i X_i$, its covariance with a standardized factor is

$$\sum_i a_i b_i \qquad [7.7]$$

where b_i is the correlation of the variable with the factor. Thus with unit weights the covariance is Σb_i since the a_i are all one. In the single-factor case the b_i can be estimated by the factor loading of X_i. To find the correlation of the factor with the composite, one must divide

Equation 7.7 by the standard deviation of the sum or composite. The variance of the composite is $\sum_i \sum_j a_i a_j r_{ij}$ where a_j is the weight for X_j, and so the correlation is

$$\frac{\Sigma a_i b_i}{(\Sigma\Sigma a_i a_j r_{ij})^{1/2}}$$

The *squared* correlation between the factor and composite is the reliability of the composite. One can then compare the reliability estimates by using different methods of forming the composites. If the weights are chosen by multiple regression, the reliability of the composite is at a maximum. Any other composite *theoretically* has lower reliability. However, the regression weighted composite does capitalize on chance, and most certainly reliability estimates are inflated. Moreover, a unit weighting of variables often correlates .95 or better with the regression weighted composite.

Cronbach's alpha is a commonly suggested measure of reliability of linear composites. It tends to underestimate the reliability of a composite as the factor loadings become more variable. It does have the advantage of computational ease. The formula for alpha using unit weights of *standardized* scores is

$$\alpha = \frac{n\bar{r}}{1 + (n-1)\bar{r}}$$

where n is the number of measures and $\bar{r}$ is average correlation between measures. The formula is sometimes called the generalized Spearman–Brown formula.

As an illustration, the reliability of factor scores is estimated for the study by Hamilton in Table 7.3. First regress F, the factor, on X_1 through X_4. The correlations among the X variables are given in Table 7.3 and the correlations with the factor are the maximum likelihood factor loadings. Obtained is the following regression estimate of the factor score using Equation 7.6:

$$3.155X_1 + .632X_2 + 1.682X_3 - 1.225X_4$$

The covariance of the preceding with F is 5.078 and its variance is 30.859. Thus the squared correlation of the composite with F is .836 which is the reliability of the regression weighted composite. If we simply sum the measures by $X_1 + X_2 + X_3 - X_4$, the variance is 9.598. The correlation of the simple sum with regression composite is .973. We have then gained little or nothing by doing the regression analysis.

The reliability of the sum is .791, which is only slightly less than .836, and so little was sacrificed by unit weights. Finally Cronbach's alpha of the sum is

$$\alpha = \frac{(4)(.4665)}{1 + (3)(.4665)}$$

$$= .778$$

which closely approximates the .791 value.

Before moving on to models with multiple unmeasured variables, which are discussed in the next chapter, some discussion about the meaningfulness of unmeasured variables is in order. Some researchers do not believe that factors have any meaning because the researcher does not know the score of each subject on the factor. Since the researcher can only derive factor scores, which do not perfectly correlate with factor, the factor remains a mysterious platonic ideal. Such a view suffers from an over-reliance on single operationalism; that is, the explanatory concepts in science are measured variables. A multiple operational approach would argue that explanatory concepts are rarely directly measured in the social sciences. Our measures of motivation, attitudes, intelligence, group cohesion, and ideology only imperfectly tap the constructs that have been posited. From this point of view measurement and unreliability go hand in hand. Thus to measure one must triangulate by using multiple operations. However, even the sum of the multiple operations does not define the construct. As is seen in the next two chapters very strong tests can be made of models with multiple operationalizations of multiple constructs. In none of the models is the construct measured, but rather the construct is simply posited to be imperfectly exemplified by its multiple operationalizations. Latent constructs are no less real than are measured variables unreal, since the factors more closely correspond to scientific concepts while measures are inherently noisy and provide only the shadows of social reality.

8

Causal Models with Multiple Unmeasured Variables

Most models with unmeasured variables usually have more than one measured variable, each with multiple indicators. The presence of multiple unmeasured variables adds many new complexities beyond a single unmeasured variable. The concern of this chapter is with what has been called the *measurement model*. It is the model that maps the measures or indicators onto the factors or unmeasured variables. The unmeasured variables may well be correlated, but in this chapter such correlations and covariances remain unanalyzed. The following chapter concerns itself with a structural analysis of these correlations.

This chapter is organized in three sections. The first section extensively examines two factor models. Most of the important material in this chapter can be found in this section. The second section considers models with three or more factors. In the third and last section the multitrait–multimethod matrix is analyzed.

TWO-CONSTRUCT MULTIPLE INDICATOR MODEL

Identification

In Figure 8.1 is a structural model with two unmeasured variables, F and G, and five measured variables, X_1 through X_5. The unmeasured

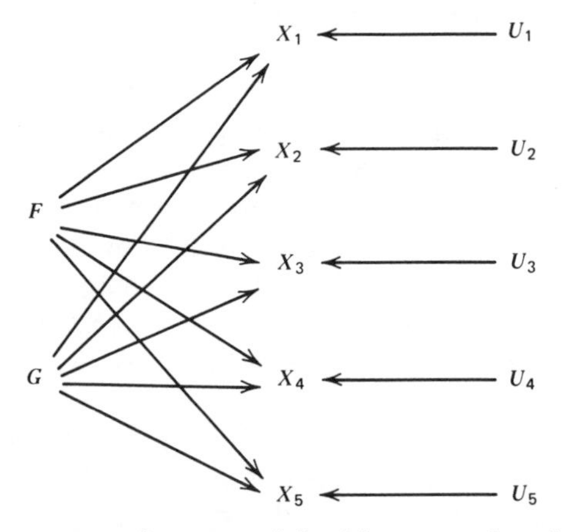

Figure 8.1 Two-factor model with uncorrelated factors.

variables or factors are assumed to be uncorrelated both with each other and with the disturbances. There are a total of 10 free parameters, the loadings on the factors, and a total of 10 correlations ($5 \times 4/2$). It would seem that the model is just-identified since there are as many parameters as correlations. Try as you can, you will never be able to solve for the parameters. It happens that a rather complex overidentifying restriction holds:

$$[\rho_{24}\rho_{35} - \rho_{25}\rho_{34}][\rho_{12}(\rho_{13}\rho_{45} - \rho_{15}\rho_{34}) - \rho_{14}(\rho_{13}\rho_{25} - \rho_{15}\rho_{23})]$$
$$- [\rho_{23}\rho_{45} - \rho_{25}\rho_{34}][\rho_{12}(\rho_{14}\rho_{35} - \rho_{15}\rho_{34}) - \rho_{13}(\rho_{14}\rho_{25} - \rho_{15}\rho_{24})] = 0$$

The reader will be happy to be spared a proof of the preceding condition, first shown by Kelley (1935). Given the preceding overidentifying restriction, one correlation is lost and there is no solution for the parameters of the model. Thus, even though the model contains an overidentifying restriction, not a single parameter is estimable. The problem cannot be solved by adding more variables. Even if there were 100 variables each loading on the two factors, it would only increase the number of overidentifying restrictions and still the parameters would remain underidentified.

To be able to identify and then estimate factor loadings for models with two or more factors, we must have constraints on the factor solution. In general, to estimate the factor loadings and factor covariance matrix, first the minimum condition of identifiability must be satisfied (that is, as many correlations as free parameters) and second p^2 parameters must be fixed where p is the number of factors. These

p^2 fixed parameters may be either factor loadings, correlations (covariances) between factors, or the variances of the factors. These parameters are usually fixed by setting their values to zero. For instance, a loading may be set to zero implying that the factor does not cause the particular measure, or a correlation (covariance) between two factors may be set to zero. A second common method of fixing a parameter is to set it to the value of one. For factor variances this implies that the factor is standardized, or for factor loadings a value of one puts the factor in the same units of measurement as the variable. Other possible methods of fixing parameters are equality and proportionality constraints.

Even when p^2 parameters are fixed the model may still not be identified since it is only a necessary but not sufficient condition for identification. Recall that the condition also holds for single factor models. Since $p^2 = 1$, there must be one constraint. It is usually either that the variance of the factor is one or one factor loading is one. Forcing the equality of two loadings is not by itself a sufficient condition to identify the model.

With two factor models there must be $2^2 = 4$ constraints. If four constraints are not made or if they are not the right four, the model is not identified. Since there is no unique solution, in factor analysis parlance the solution can be *rotated*. Rotation and all its attendant problems are brought about because most factor analysis methods do not specify an identified model. The square root method of factor analysis does presume an identified model (Harmon, 1967, pp. 101–103). It constrains the p factor variances to be one (p constraints), the factor correlations to be zero ($p(p-1)/2$ constraints), and for the loading matrix to be triangular, that is, all variables load on the first factor, all but the first variable load on the second factor, and so on until only the last measure loads on the last factor ($p(p-1)/2$ constraints). Such a model may be identified with a total of p^2 constraints. However, to satisfy the minimum condition of identifiability, a loading of each factor must be fixed. Although it may be identified it rarely makes any structural sense.

The approach taken in this chapter is to allow each measure to load on only one factor. In most cases this will lead to identification, as well as allow the factors to be correlated.

The Model

Consider a two-factor model with each factor having four indicators as in Figure 8.2. Four measures, X_1 through X_4, are *indicators* of F and the

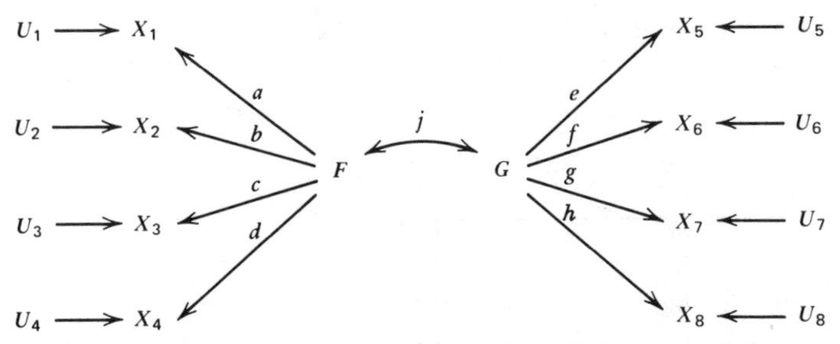

Figure 8.2 Two-factor multiple indicator model.

other four, X_5 through X_8, are indicators of G. The two factors are correlated, or oblique. In Figure 8.2 it is assumed that indicators of one construct are not caused by the other construct and that the unique factors, or disturbances, are uncorrelated with each other and with the two constructs. An *indicator* then is defined as a measure that is caused by a single construct. Thus, in this case the indicators of G do not load on F and the indicators of F do not load on G. The factor loading matrix is:

	F	G
X_1	a	0
X_2	b	0
X_3	c	0
X_4	d	0
X_5	0	e
X_6	0	f
X_7	0	g
X_8	0	h

As in the previous chapter, the correlation of the indicator with the construct is called the *epistemic correlation*. In Table 8.1 the tracing rule has been used to derive the theoretical correlations for the model in Figure 8.2. The correlations in the box of Table 8.1 are called the *cross-correlations*. These correlations involve the correlations between indicators of different constructs. The correlations above the box are correlations of the indicators of F, and beside the box are the correlations of the indicators of G.

Table 8.1. Theoretical Correlations for Figure 8.2

X_1								
X_2	ab							
X_3	ac	bc						
X_4	ad	bd	cd					
X_5	aej	bej	cej	dej				
X_6	afj	bfj	cfj	dfj	ef			
X_7	agj	bgj	cgj	dgj	eg	fg		
X_8	ahj	bhj	chj	dhj	eh	fh	gh	
	X_1	X_2	X_3	X_4	X_5	X_6	X_7	X_8

The minimum condition of identifiability has been established for the model in Figure 8.2 since there are 28 correlations between measured variables, and only 9 free parameters, 8 factor loadings, and 1 factor correlation. Since 8 factor loadings have been constrained to be zero and the variance of the factors to be 1, enough constraints have been made to obtain a solution. Not only is the model identified, there is also a plethora of overidentifying restrictions, 19 in all (28 − 9). If the researcher chooses to set the units of the factor to that of a measure by fixing the measure's loading to one, then there are still 19 overidentifying restrictions. In such a case the loading is no longer free, but the factor variance is free.

Estimation

The traditional factor analytic solution to the model in Figure 8.2 would be a principal components solution with some form of oblique rotation. However, unless all the overidentifying restrictions are perfectly met, no rotation will force the loadings to be zero where they should be. Some readers may have been faced with the difficult decision of how oblique should an oblimax rotation be. A Procustes rotation (Gorsuch, 1974, pp. 166–168) might be tried, but it requires not only the specification of the zero loadings but also exact values of the nonzero loadings. Another factor analytic solution to the model in Figure 8.2 is the *multiple group solution* (Harmon, 1967; Gorsuch, 1974) which is rarely used. Like most factor analytic solutions, it too suffers from the communality problem; that is, communalities must be specified in advance. A variant of the multiple group solution to the model in Figure 8.2 is cluster analysis (Tyron & Bailey, 1970).

The path analytic estimates which were originally proposed by Costner (1969) are straightforward. To estimate the epistemic correlations simply pool all the estimates of each. An estimate of the epistemic correlation for X_i can be shown to be

$$\left(\frac{r_{ij}r_{ik}}{r_{jk}}\right)^{1/2}$$

where X_j is an indicator of the same construct as X_i whereas X_k need not be. Then to pool the estimates of the squared epistemic correlation, sum the absolute values of the numerators of all estimates and divide it by the sum of the absolute values of the denominators. If there are a indicators of construct F and b of G, there are a total $(a - 1)(b + a/2 - 1)$ estimates of the epistemic correlation for each indicator of G and $(b - 1)(a + b/2 - 1)$ for each indicator of F.

To estimate ρ_{FG} pool all of its estimates. They are of the form

$$r_{FG}^2 = \frac{r_{ik}r_{jl}}{r_{ij}r_{lk}}$$

where i and j are indicators of F and l and k indicators of G. There are a total of $ab(a - 1)(b - 1)/2$ possible estimates to be pooled. Again the numerators, and then the denominators, of each estimate are summed. The estimate is the ratio of the pooled numerator to the pooled denominator.

The only estimation procedure based on a statistical analysis is maximum likelihood factor analysis (Jöreskog, Gruvaeus, & van Thillo, 1970). A two-factor model is specified, with each variable loading on a single factor. The two factors can be correlated. The ACOVS or LISREL programs can allow correlated disturbances, equal standardized or unstandardized loadings, zero-factor correlations, or equal error variances. The overall fit of the model can be evaluated and restricted models can be compared by χ^2 goodness of fit tests. All the examples are estimated by employing maximum likelihood factor analysis.

Specification Error

Instead of testing whether *all* the overidentifying restrictions hold for the model in Figure 8.2 simultaneously, it is useful to partition the restrictions into tests of different constraints on the model.

One obvious set of restrictions to examine is the correlations within a construct. The correlations within each construct should evidence a single-factor structure. For factor F, since there are four measures, there are two overidentifying restrictions. These are referred to as the test of *homogeneity within indicators of a construct.* More compactly, the restrictions are called *homogeneity within.* As shown in the previous chapter, homogeneity within implies tetrads of the form $\rho_{ij}\rho_{kl} - \rho_{ik}\rho_{jl} = 0$, where i, j, k, and l are all indicators of the same construct.

Homogeneity between constructs can also be defined. An examination of the correlations in Table 8.1 between indicators of F and G (the cross-correlations), evidences a number of overidentifying restrictions. The restrictions follow from the fact that the cross-correlations can be used to estimate the ratio of epistemic correlations. For instance, to estimate a/b one could examine

$$\frac{a}{b} = \frac{\rho_{15}}{\rho_{25}} = \frac{\rho_{16}}{\rho_{26}}$$

which implies a vanishing tetrad of

$$\rho_{15}\rho_{26} - \rho_{16}\rho_{25} = 0$$

In general, denoting i and j as indicators of F, and k and l as indicators of G, the following vanishing tetrad holds:

$$\rho_{ik}\rho_{jl} - \rho_{il}\rho_{jk} = 0$$

The final set of restrictions is called the test of *consistency of the epistemic correlations.* From the correlations within a construct, the epistemic correlations can be estimated. From the cross-correlations the ratio of these epistemic correlations can also be estimated. These two sets of estimates should be consistent. A test of consistency is again a vanishing tetrad since

$$\frac{a}{b} = \frac{\rho_{13}}{\rho_{23}} = \frac{\rho_{15}}{\rho_{25}}$$

which implies

$$\rho_{13}\rho_{25} - \rho_{23}\rho_{15} = 0$$

Since from the ratios a/b, b/c, and c/d all the remaining ratios can be obtained, there are then a total of three restrictions on the consistency of the estimates of each construct. In general consistency of F implies a tetrad of the form

$$\rho_{ij}\rho_{kl} - \rho_{ik}\rho_{jl} = 0$$

where i, j, and k are indicators of F and l an indicator of G.

If it is assumed that there are a indicators of construct F and b of G, the total number of correlations is $(a + b)(a + b - 1)/2$ and the total number of free parameters in the model is $a + b + 1$. Subtracting the latter from the former yields $(a + b)(a + b - 3)/2 - 1$, which is the total number of overidentifying restrictions on the model. These restrictions can be divided into homogeneity and consistency restrictions. There are a total of $(a^2 - a)(a - 2)(a - 3)/8$ tetrads testing homogeneity within construct F of which only $a(a - 3)/2$ are independent. Similarly for construct G, there are $(b^2 - b)(b - 2)(b - 3)/8$ tetrads with $b(b - 3)/2$ independent restrictions. For homogeneity between constructs, there are a total of $ab(a - 1)(b - 1)/4$ tetrads of which $(a - 1)(b - 1)$ are independent. For the consistency of F, there are a total of $(a^2 - a)(a - 2)b/2$ tetrads, but only $a - 1$ are independent given homogeneity within construct F and homogeneity between constructs F and G. Similarly for the consistency of construct G, there are $(b^2 - b)(b - 2)a/2$ tetrads of which only $b - 1$ are independent. If one adds up the total number of independent restrictions they equal $(a + b)(a + b - 3)/2 - 1$, which is the number of overidentifying restrictions on the entire model.

To assess homogeneity within a construct one needs at least four indicators of the construct. For homogeneity between, one needs only two indicators for both constructs. To test for consistency, one needs three indicators for the construct whose consistency is being tested and only one indicator of another construct.

To statistically evaluate the different constraints on the model, I suggest the following procedure. I am sure that there are other procedures that are better, but this method is fairly straightforward and simple. To test for homogeneity within a construct, test for a single factor by using maximum likelihood factor analysis. If the indicators of the factor are not single-factored, test each of the tetrads using canonical correlation. Hopefully, an examination of the tetrads that fail to vanish may point to the troublesome indicators. To test for homogeneity between, perform a canonical correlation analysis treating the indicators of one construct as one set and the indicators of the other construct as the second set. The test that the second canonical correlation is zero will test for homogeneity between constructs. If the test fails then the individual tetrads can be tested to point out the troublesome indicators.

There is no simple test for consistency. A rather tedious strategy is to test all the tetrads. This should be done only after both the homogeneity between and within tests have been passed.

If a given test fails, what sort of specification error is indicated? If homogeneity within is not satisfied, it indicates that the disturbances of the measures are correlated or there may be a second factor. If homogeneity between is not satisfied, most likely disturbances are correlated across constructs. For instance, two indicators may share method variance.

The specification error for consistency is rather interesting. For Figure 8.2 imagine there is a path from the construct G to X_1 and thus X_1 would load on both F and G. In such a case homogeneity within and between would be satisfied as well as the consistency of the indicators of G, but the indicators of F would not evidence consistency. In particular, tetrads involving the correlations of X_1 with an indicator of F and an indicator of G would not vanish. So in general, violations of homogeneity within indicate correlated errors within indicators of a construct, violations of homogeneity between indicate disturbances correlated across different constructs, and violations of consistency indicate that certain measures load on both factors. Of course, if the model is grossly misspecified such tests may not be particularly diagnostic. However, a model with a few specification errors can be improved by such an analysis.

Identification Revisited

The reader is probably wondering what is the minimum number of indicators necessary for multiple indicator models. Since obtaining indicators of constructs may be difficult, it would be helpful to know the smallest number that are needed. Consider the model in Figure 8.3. There are only two indicators of each of the two constructs, four variables in all. The model is, however, identified. For instance, the parameter a^2 equals:

$$a^2 = \frac{\rho_{12}\rho_{13}}{\rho_{23}} = \frac{\rho_{12}\rho_{14}}{\rho_{24}}$$

Note, however, that the denominators of both the preceding formulas depend on ρ_{FG}:$\rho_{23} = bc\rho_{FG}$ and $\rho_{24} = bd\rho_{FG}$. Thus, if ρ_{FG} equals zero or

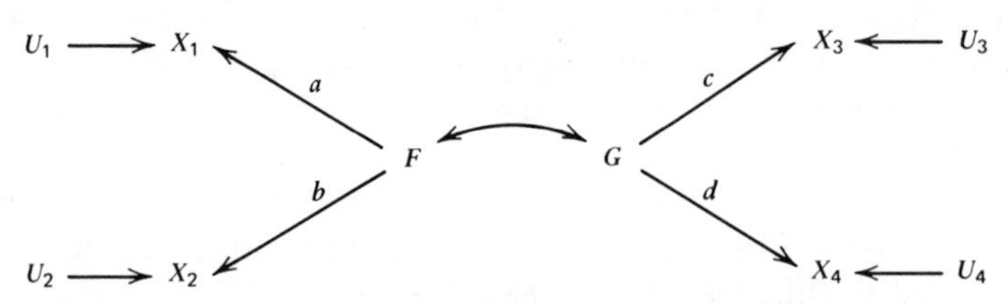

Figure 8.3 Two-factor model, each factor with two indicators.

nearly equals zero, the model is empirically underidentified since the denominators are zero or near zero. Thus, assuming no correlated disturbances, it is possible to estimate loadings given that the construct correlates with other constructs in the model. Of course, as in single-factor models the standardized loadings must not be small or else the model is again empirically underidentified.

Although two indicators are sufficient to allow for identification, it is desirable to have at least three indicators per construct. One reason is fairly obvious. The model may be identified in principle, but under-identified empirically, that is, the constructs may correlate low or the loadings may be too small. A second problem is specification error. The investigator may be unable to allow for correlated disturbances if there are only two indicators per construct. Consider the model in Figure 8.5, which is discussed later in this chapter. Note that the disturbances of X_1 and X_4, X_2 and X_5, and X_3 and X_6 are correlated. Such a model may be very plausible, if the X_1 and X_4 are the same variable measured at times 1 and 2, respectively, and X_2 and X_5 are again the same variable measured at two points in time as well as X_3 and X_6. The model in Figure 8.5 is identified with three indicators at each time point; it would not be if there were only two indicators.

Just as three indicators are better than two, so are four better than three. With only three indicators it is not possible to test for homogeneity within a construct. With four such a test can be made and possible correlated errors within indicators of a construct can be spotted. Having five indicators is only slightly better than having four. I suggest the following rule of thumb: Two *might* be fine, three is better, four is best, and anything more is gravy. To complicate matters slightly in some very special circumstances, it is seen in the next chapter that a construct need have only a single indicator.

Illustrations

In Table 8.2 there are a set of correlations taken from data gathered by Brookover and reanalyzed by Calsyn and Kenny (1977). The measures were taken from 556 white eighth-grade students and are:

X_1—self-concept of ability

X_2—perceived parental evaluation

X_3—perceived teacher evaluation

X_4—perceived friends' evaluation

Table 8.2. Correlations between Ability and Plans below Diagonal Residuals above Diagonal[a,b]

	1	2	3	4	5	6
X_1	1.000	−.003	.005	−.020	.014	.026
X_2	.730	1.000	−.004	.020	−.009	−.006
X_3	.700	.680	1.000	.010	−.016	−.018
X_4	.580	.610	.570	1.000	.011	−.020
X_5	.460	.430	.400	.370	1.000	.000
X_6	.560	.520	.480	.410	.720	1.000

[a]Data taken from Calsyn and Kenny (1977) and correlations are computed to only two decimal places.
[b]N = 556.

X_5—educational aspiration

X_6—college plans

The three perceived measures were all the perceptions of the student. In Figure 8.4 there is a model for the six variables. The four measures of evaluation of the child are assumed to be indicators of self-perception of ability, A, and the two aspiration measures are called plans, or P. Also in Figure 8.4 there are the parameter estimates of the model using maximum likelihood factor analysis. The epistemic correlations are all high, as is the correlation between A and P. It is instructive to note that *none* of the cross-correlations are as large as the estimate of the true correlation between constructs since all the cross-correlations are attenuated because of measurement error in the indicators. It is instructive to compare alternative strategies of estimating the correlation between A and P. If one randomly picked a single indicator of A and correlated it with a randomly chosen indicator of P, the correlation between A and P would be estimated as .454. If one unit weighted (added up the standard scores) both sets of indicators, the correlation would be .571. If one computed a canonical correlation between the two sets of indicators, it would be .597. Even though the reliabilities of the indicators are moderate to large, all the estimates by standard procedures are appreciably lower than the .666 value estimated by confirmatory factor analysis.

The degrees of freedom for this model are $(6)(3/2) - 1 = 8$. The overall χ^2 is 9.26 which is well within the limits of sampling error. In fact, as can be seen by looking above the diagonal in Table 8.4, the

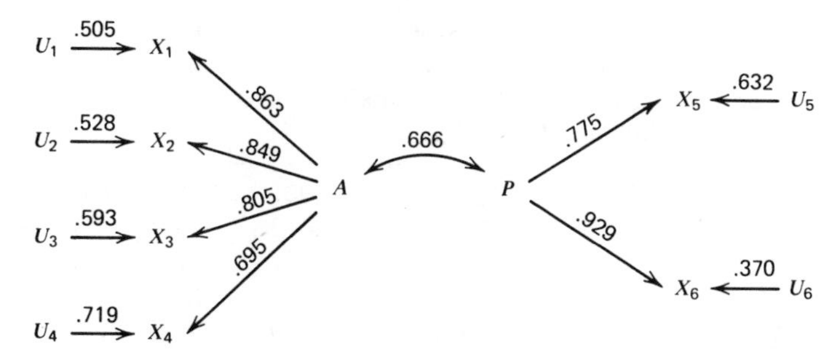

Figure 8.4 ***Multiple indicator model for the correlations in Table 8.2.***

largest residual is only .026. The power of the goodness of fit test should be high since the sample size and the correlations are large. To obtain a χ^2 test for the homogeneity of the indicators of A, a test of single-factoredness of the four indicators of A is conducted. A χ^2 of 3.13 is obtained with $(4)(1/2) = 2$ degrees of freedom. This χ^2 is not significant and so single factoredness of the A indicators is indicated. Since there are only two indicators of the P factor, one cannot test for homogeneity within it. The test of homogeneity between constructs A and P can be performed by testing whether the second canonical correlation between indicators of A and of P is zero. The χ^2 equals 1.95 with 3 degrees of freedom which is clearly not significant. To test for consistency, tests of the 24 tetrads were performed. Since there are multiple tests alpha (.05) was divided by 24, the number of tests. All the tetrads vanished using this significance level.

The second illustration is taken from a secondary analysis conducted by Milburn (1978). One hundred and forty adult smokers were asked questions on a four-point scale about smoking:

X_1: unbearable to run out of cigarettes

X_2: impossible to cut down

X_3: feel more relaxed after smoking

The X_4 to X_6 measures are the same with the subjects being measured one year later. All subjects were smokers from a town in northern California. In Table 8.3 are the correlations between items and the standard deviation of each item. All the solutions reported below are unstandardized since the loadings of X_1 and X_4 are forced to one, and the variance-covariance matrix serves as input. The variance of the time 1 and time 2 factor is then free to vary.

Table 8.3. Correlations of Cigarette Smoking Items at Two Time Points, X_1 to X_3—Time 1 and X_4 to X_6—Time 2. Residuals on and above Diagonal[a,b]

X_1	.012	.038	−.010	.009	−.070	.019
X_2	.445	.016	−.017	−.018	−.040	.022
X_3	.274	.302	−.012	.122	−.041	.020
X_4	.601	.268	.340	−.002	−.040	.026
X_5	.201	.340	.162	.270	−.031	.042
X_6	.222	.249	.484	.268	.293	.022
Standard deviation	1.053	1.097	.909	.937	1.055	.800

[a]N = 140.
[b]Data taken from Milburn (1978).

The first solution attempted forced the loadings of first, X_2 and X_5, and second, X_3 and X_6, to be equal. Thus the unstandardized loadings were constrained to be stable over time. The χ^2 for the model is 47.19 with 10 degrees of freedom. Clearly some part of the model is misspecified. Since the same items were used at both points in time it is likely that the unique factors are correlated over time. To test this a canonical correlation analysis was set up treating X_1, X_2, and X_3 as one group of variables and the remaining variables as the second set. The test of the $\chi^2(4)$ equals 29.72, which is highly significant. Thus correlated errors across constructs is indicated.

The model was respecified allowing the disturbances of the same measure to be correlated over time. The variance of the disturbances was set at one. The resultant parameter estimates are contained in Figure 8.5. The fit of the model is quite good: $\chi^2(7) = 10.30$. Residuals from the *covariance* matrix are reproduced in Table 8.3. Interestingly, the factor variance decreases over time being .406 at time 1 and .262 at time 2. Respondents seem to be becoming more homogeneous over time. The autocorrelation of the factor is $.251/((.406)(.262))^{1/2} = .770$.

To test the equality of the loadings over time, a model was set up that allowed the loadings to vary over time. However, to identify the model, the loadings for X_1 and X_4 were kept at one. The χ^2 for this model is 9.44 with 5 degrees of freedom. The test of equality constraints is the difference of the general model from the restricted model: $\chi^2(7)$ of 10.30 minus $\chi^2(5)$ of 9.44 or $\chi(2)$ of .86, which is clearly not significant, indicating that the fit does not suffer by imposing the equality constraints.

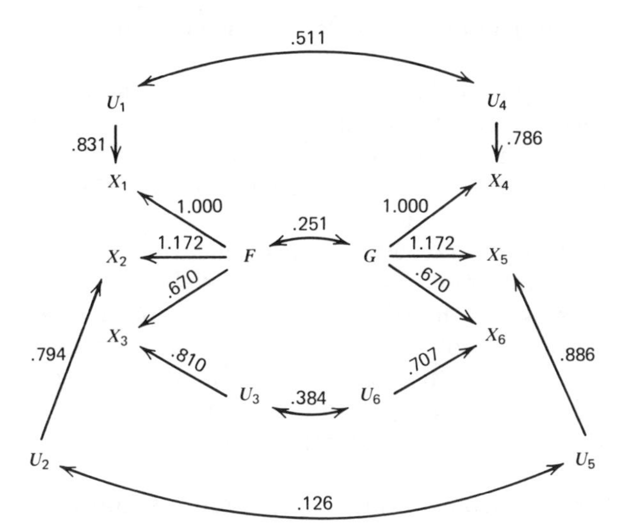

Figure 8.5 Unstandardized model for the covariance matrix in Table 8.3.

THREE-OR-MORE-CONSTRUCT MULTIPLE INDICATOR MODEL

Complications

The addition of a third construct to multiple indicator models adds only a few complications. As the number of constructs grows the number of correlations (covariances) between the constructs increases geometrically. The researcher may desire to specify a structural model for those relationships. Such models are discussed in the next chapter.

Multiple indicator models with many constructs usually tend to be highly overidentified. For instance, a model with four constructs, each with four indicators, with each indicator loading on only one factor, and no correlated errors has 98 overidentifying restrictions! For such models a small number of specification errors can greatly distort the parameter estimates and lack of fit can be virtually undetectable if one solely relies on the overall χ^2 test. The problem is analogous to an overall F test in the analysis of variance. Since there are so many degrees of freedom on the numerator, the overall test has low power to detect a few but important differences. Clearly one must think through the model and explicitly test for specific sorts of specification error.

Failure to do so may often lead to parameter estimates that are very misleading.

It is not unusual for the structural model to contain a mix of constructs. Some of the constructs require multiple indicators while others may not. For instance, the model may contain demographic variables such as sex, ethnicity, or age. It seems reasonable to argue that these variables have perfect or near perfect reliabilities: They are perfect indicators with epistemic correlations of one. How can such variables be incorporated into multiple indicator models?

The simplest way to do so is to create a factor for each such measure. Then force the loading of the measure on the construct to be one and do not give the measure a disturbance. The factor will then become the measure, and its variance is set to the variance of the measure. Thus, if the variable is already standardized, the factor will also be standardized.

Before turning to an example of a multiple construct model, it should be noted that specification error can be evaluated in a different manner with three or more constructs. Earlier, consistency of estimates was defined. It involved comparing the ratio estimates of loadings determined by the *within* and *between* construct correlations. With three or more constructs consistency can be evaluated by comparing two sets of *between* construct correlations. For instance, note that in Figure 8.6 the ratio of the path from L to X_1 to the path from L to X_2 equals

$$\frac{\rho_{13}}{\rho_{23}} = \frac{\rho_{14}}{\rho_{24}} = \frac{\rho_{15}}{\rho_{25}} = \frac{\rho_{16}}{\rho_{26}}$$

The preceding yields six vanishing tetrads only three of which are independent:

$$\rho_{13}\,\rho_{24} - \rho_{23}\,\rho_{14} = 0 \qquad [8.1]$$

$$\rho_{13}\,\rho_{25} - \rho_{23}\,\rho_{15} = 0$$

$$\rho_{13}\,\rho_{26} - \rho_{23}\,\rho_{16} = 0$$

$$\rho_{14}\,\rho_{25} - \rho_{24}\,\rho_{15} = 0$$

$$\rho_{14}\,\rho_{26} - \rho_{24}\,\rho_{16} = 0$$

and

$$\rho_{15}\,\rho_{26} - \rho_{25}\,\rho_{16} = 0 \qquad [8.2]$$

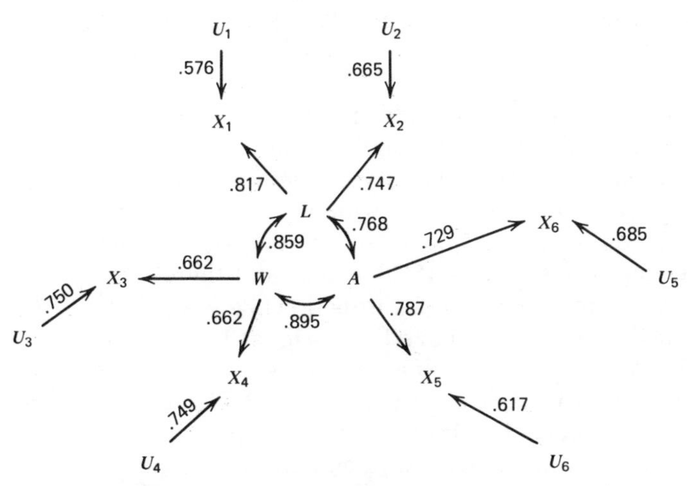

Figure 8.6 Multiple indicator model for the correlations in Table 8.4.

Two of the overidentifying restrictions are constraints of homogeneity first between L and A (Equation 8.1) and second between L and W (Equation 8.2). The final independent restriction can be called the consistency of L with W and A. Only when a construct has just *two* indicators does it make sense to evaluate consistency with two other constructs. If there are three or more indicators this form of consistency follows from both simple consistency and homogeneity between constructs. In a similar fashion the consistency of A with L and W and the consistency of W with L and A can be defined.

In Table 8.4 are the correlations between six subtests of the Iowa Test of Basic Skills for 1501 inner-city children (Crano, Kenny, & Campbell, 1972). The six tests can be thought of as two indicators of language

Table 8.4. Correlation between Six Achievement Tests below the Diagonal and Residuals above[a,b]

X_1	1.0000	−.0000	−.0060	.0030	−.0130	.0200
X_2	.6108	1.0000	.0130	−.0090	.0080	−.0150
X_3	.4591	.4373	1.0000	.0000	.0040	−.0090
X_4	.4679	.4164	.4384	1.0000	.0030	−.0000
X_5	.4816	.4597	.4701	.4694	1.0000	−.0000
X_6	.4782	.4030	.4225	.4316	.5736	1.0000
	X_1	X_2	X_3	X_4	X_5	X_6

[a] N = 1501.
[b] Data taken from Crano, Kenny, & Campbell (1972).

usage (capitalization, X_1, and punctuation, X_2), work skills (map reading, X_3, and use of graphs, X_4), and arithmetic (concepts, X_5, and problems, X_6).

In Figure 8.6 is a path model and the parameter estimates from the Jöreskog program. The work skill measures have the lowest loadings. Not surprisingly the correlations among the factors are very high. The $\chi^2(6)$ goodness of fit test is 9.98, which is not significant at conventional levels of significance. There are six degrees of freedom since the number of correlations is 15 ($6 \times 5/2$) and there are 9 parameters: 6 epistemic correlations, and 3 correlations between factors. The residuals from the model are above the diagonal in Table 8.4.

It is impossible to test for homogeneity within a construct since there are only two indicators per construct. However, a test for homogeneity between each pair of constructs is possible. The test of homogeneity between language usage and work skills is

$$\rho_{13}\rho_{24} - \rho_{14}\rho_{23} = 0$$

between language usage and arithmetic

$$\rho_{15}\rho_{26} - \rho_{16}\rho_{25} = 0$$

and between work skills and arithmetic

$$\rho_{35}\rho_{46} - \rho_{36}\rho_{45} = 0$$

The χ^2 tests of the second canonical correlation for each tetrad are 1.58, 7.06, and .17, respectively. Obviously there is not homogeneity between language usage and arithmetic whereas there is homogeneity between the other two pairs.

To test for consistency we compute all the tetrads of the form

$$\rho_{ij}\rho_{kl} - \rho_{ik}\rho_{jl}$$

where i and l are the indicators of one construct, j an indicator of a second construct, and k an indicator of a third construct. There are a total of 12 such tetrads. Using .05/12 as the alpha level, all the 12 such tetrads vanish.

MULTITRAIT–MULTIMETHOD MATRIX

Campbell and Fiske (1959) suggested the multitrait–multimethod matrix as a means for assessing validity. This matrix is a correlation matrix between the traits or constructs, each of which is measured by the same set of methods. In a sense, the measures are formed by factorially combining traits and methods. The resulting correlations can be characterized as same-trait, different-method correlations, different-trait, same-method correlations, and different-trait, different-method correlations. According to Campbell and Fiske (1959) convergent validity is indicated if the same-trait, different-method correlations are large. Discriminant validity is indicated if those same correlations are larger than the different-trait, different-method correlations. Method variance is indicated if the different-trait, same-method correlations are larger than the different-trait, different-method correlations.

These rules of thumb are useful as far as they go, but ideally the matrix should be analyzed more formally. Numerous procedures have been put forward to factor analyze the matrix, two of which are considered below. There is no single way to specify a factor model.

The first is called the classical test theory formalization. The traits are factors whereas the disturbances or unique factors are allowed to be correlated across measures using the same method. Such a model is identified if there are at least two traits and three methods. Assuming the model fits the data, then convergent validation is assessed by high loadings on the trait factors, discriminant validation by low to moderate correlation between the trait factors, and method variance by highly correlated disturbances.

The second formalization was originally suggested by Jöreskog (1971) and has been extensively reviewed by Alwin (1974). Here each method and trait are taken to be factors and the factors may be intercorrelated. Consequently, each variable loads on both a trait and method factor. Discriminant and convergent validation are assessed as with the previous formalization, but method variance is now indicated by the loadings on the method factors. For such a model to be identified, at least three traits and three methods must be measured.

This second formalization should, in general, be preferred over the first. The first presumes that method variance is uncorrelated with both traits and the other methods. The second procedure does, however, assume that the effect of a method is single-factored whereas the first procedure does allow for multidimensional effects.

Illustrations

Jaccard, Weber, and Lundmark (1975) measured two constructs, attitudes toward cigarette smoking (C) and capital punishment (P), by four different methods: semantic differential (method 1), Likert (method 2), Thurstone (method 3), and Guilford (method 4) methods.

Table 8.5. Correlations of Four Attitude Measures with Residuals on and above the Diagonal[a,b]

C_1	.001	.019	−.006	−.008	.059	.068	.057	.056
C_2	.780	−.010	−.009	−.023	.046	.047	.116	.036
C_3	.810	.770	.005	.024	−.007	−.037	−.002	.011
C_4	.760	.710	.810	.004	−.085	−.114	−.116	−.101
P_1	.290	.230	.190	.100	.021	.011	.016	.004
P_2	.280	.290	.180	.090	.840	.016	.014	−.012
P_3	.260	.310	.240	.080	.810	.890	−.006	−.033
P_4	.270	.240	.230	.150	.840	.910	.850	−.037
	C_1	C_2	C_3	C_4	P_1	P_2	P_3	P_4

[a]N = 35.
[b]Data taken from Jaccard et al. (1975).

The sample size was only 35. In Table 8.5 are the correlations between the eight variables. Convergent validation is indicated because the measures of the same construct correlate highly. Discriminant validation is also indicated because the cross-construct correlations are all smaller than the within-construct correlations. There is some indication of systematic method variance since the correlations between different constructs using the same method are larger than the correlations between measures of different constructs and different methods.

Consider the path diagram in Figure 8.7 for the Jaccard et al. study. In this figure there are two types of variables, measured and unmeasured. The two traits, C = cigarette smoking and P = capital punishment, are the common factors and their four measures load on each. None of the cigarette smoking measures load on the capital punishment factor or vice versa. The factors designated E_1 through E_8 are the unique factors, or disturbances, and one and only one measure loads on each. These unique factors are assumed to be uncorrelated with the trait factors and with the other unique factors with the important exception that unique factors may be correlated across measures using the same method. This exception allows for the method effects.

The minimum condition of identifiability is met since the number of correlations, 28, exceeds the number of parameters. There are 13 parameters, 8 factor loadings, and 5 factor correlations. The difference is 15, which is the total number of independent overidentifying restrictions on the model in Figure 8.7.

The values in Figure 8.7 are the parameter estimates. Note that the loadings on each trait factor are high and fairly homogeneous. There is a small positive correlation between the cigarette smoking and advocacy

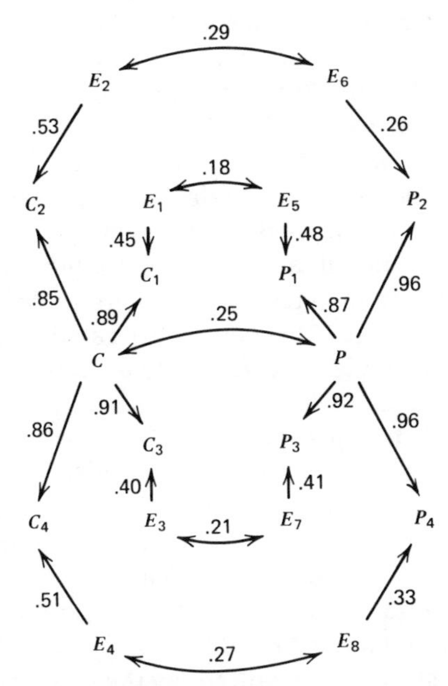

Figure 8.7 ***Multiple indicator model for correlations in Table 8.5.***

of capital punishment factors. There are also only small, positive correlations between disturbances.

The $\chi^2(15)$ test of goodness of fit yields a value of 10.35. Since the χ^2 is not statistically significant, the parameters of the model fit the correlations. Given the low sample size, most certainly other models fit the data.

Discriminant validation implies that the correlation between traits is low. If both traits were identical, the correlation between trait factors would be near one. One can force ACOVS to make the correlation between traits to be one. (Making the correlation between factors one is equivalent to a single-factor model.) The resultant $\chi^2(16)$ for that model equals 120.75. To test for discriminant validation simply subtract this χ^2 with the earlier one and subtract their degrees of freedom to obtain a $\chi^2(1)$ of 110.40. Since this value is highly significant, the assumption that the trait correlation is unity is not confirmed and discriminant validation is indicated.

In a similar fashion one can test for the equality of the factor loadings and factor correlations. To test whether the measures of C have significantly different loadings, constrain them to be equal. The χ^2 test of such a model is 10.71 with 18 degrees of freedom. Three degrees of freedom are added over 15 since only one factor loading need be estimated, not four. Again subtracting χ^2 values, the χ^2 test of equal loadings on C is .36 with 3 degrees of freedom. A similar result is obtained for the measures of P: $\chi^2(3) = 1.17$. Thus, there is no significant difference between methods in tapping the underlying traits. A similar test of the equality of correlated errors yields another nonsignificant χ^2 of .16 with 3 degrees of freedom.

The second example of a structural analysis of the multitrait–multimethod matrix is taken from Rosengren, Windahl, Hakansson, and Johnsson-Smargdi (1976). They investigated three different scales designed to measure style of viewing television: parasocial viewing (P), capture (C), and identification (I). Three hundred and eighty-six Swedish adolescents completed the three scales for television in general (method 1), a parasocial show (method 2), and an identification show (method 3). The resultant nine scales were intercorrelated, and are presented in Table 8.6. Since there are three traits and three methods, each measure can load on each trait and each method factor. Such a solution was attempted but it yielded an anomalous result. The $\chi^2(3)$ is less than one and the largest residual is only .009. The fit is excellent but the parameter estimates are odd to say the least. The loadings on the identification factor are less than .2 for two of the scales and one is even negative. The correlation of this factor with two other factors is larger than one. A loading on another factor is larger than one, an implausible if not impossible value. It was judged that the loadings on this factor are empirically underidentified. Moreover, it seems the data are overfit, that is, that too many parameters are estimated. Evidence for overfit is

Table 8.6. Television Style Scales, Correlations below Diagonal and Residuals above [a,b]

P_1	1.000	.007	−.017	.025	.023	−.006	−.005	−.031	.001
C_1	.640	1.000	.007	.005	−.012	−.007	.003	−.012	−.009
I_1	.490	.550	1.000	−.012	.020	−.044	−.005	.030	.016
P_2	.540	.440	.250	1.000	−.003	.002	−.013	.004	.007
C_2	.330	.350	.280	.640	1.000	.001	.008	.024	.012
I_2	.240	.280	.260	.540	.700	1.000	−.036	−.032	−.023
P_3	.680	.570	.430	.520	.330	.220	1.000	.008	.001
C_3	.520	.610	.510	.440	.430	.300	.670	1.000	−.001
I_3	.430	.460	.660	.250	.310	.350	.530	.690	1.000
	P_1	C_1	I_1	P_2	C_2	I_2	P_3	C_3	I_3

[a] N = 382.
[b] Data taken from Rosengren et al. (1976).

Table 8.7. Loading and Correlation Matrix for the Correlation Matrix in Table 8.6[a]

	P	C	1	2	3
P_1	.574		.593		
C_1		.518	.844		
I_1			.803		
P_2	.531			.630	
C_2		.241		.869	
I_2				.822	
P_3	.683				.601
C_3		.627			.970
I_3					.972

	P	C	1	2	3
P	1.000				
C	.673	1.000			
1	.066	−.324	1.000		
2	.046	−.079	.461	1.000	
3	−.083	−.412	.825	.467	1.000
	P	C	1	2	3

[a]Note the I factor is omitted because of empirical underidentification.

that χ^2 is smaller than its degrees of freedom and the anomalous estimates.

Since the identification scale has only one variable with a large loading, the factor cannot be distinguished from the disturbance and the loadings are empirically underidentified. It was decided the identification factor should be dropped. This left two trait factors and three method factors. The resultant solution did not fit nearly as well: $\chi^2(11) = 19.28$. The residuals presented above the diagonal in Table 8.6 are not large, however. The loadings and factor correlations in Table 8.7 indicate more method variance than trait variance and a rather high correlation between the two traits. (No path diagram is presented since it would be too cluttered to be interpretable.) Employing the criteria offered by Campbell and Fiske, Rosengren et al. arrived at very different conclusions.

In the next chapter the correlations among the factors are analyzed. It is seen that one can impose on unmeasured variables the same type of models that are imposed on measured variables in Chapters 4, 5, and 6.

9

Causal Models with Unmeasured Variables

The preceding two chapters have exclusively discussed the question of multiple indicators of latent constructs. What has been developed is called the measurement model. This chapter is devoted to causal models in which the latent variables are endogenous and exogenous; that is, there is a causal model for the unmeasured variables. In the previous chapter it was shown how correlations between unmeasured variables can be estimated. Here these correlations are analyzed. The chapter is divided into two major sections. The first section elaborates the issues for structural models with latent endogenous and exogenous variables. The second and larger section works through five examples which were chosen to illustrate the broad range of models amenable to this type of analysis.

GENERAL CONSIDERATIONS

Measurement Model

It is helpful to employ matrices to summarize the measurement and causal models. Unfortunately the term matrix strikes fear in the heart of too many researchers. The purpose of matrices here is just the same as that of a path diagram. A matrix conveniently summarizes a set of structural equations. For the measurement model there is a matrix of measures by constructs. The row variables are all the measured variables while the column variables are the unmeasured variables. A nonzero entry in the matrix denotes that the measured variable is caused by the construct or, to put it differently, the measure is an indicator of the construct. Normally there is the specification that certain entries are zero. If this matrix seems to be similar to a factor loading matrix, it is no accident since the constructs are factors.

It is helpful to make a distinction between a measure that loads on a single factor and one that loads on two or more factors. Strictly speaking a measure *indicates* a construct only if the measure has only one nonzero entry, that is, loads on only one factor. In contrast, an example of a measure with multiple entries can be taken from the Jöreskog (1971) model for the multitrait–multimethod matrix discussed in the previous chapter. There each measure is assumed to be caused by both a method and a trait factor.

Normally each of the measures is not totally determined by the latent constructs or common factors. A measure has a disturbance or a unique factor. In some special cases a *latent* factor is defined by a measure which means the construct has only a single indicator and that measure has no disturbance. For instance, if sex were one of the measured variables, it would have no disturbance and the construct on which sex loads is then defined as sex. In the main, however, a disturbance is needed for each measure.

These disturbances are usually assumed to be uncorrelated with the constructs. They need not, however, be uncorrelated with each other. It may be that measurement errors are correlated across measures, and a covariance matrix for the disturbances is then needed. The diagonal values of that matrix would be the variances and the off-diagonal values would be the covariances.

The disturbances can be viewed as measurement errors given a special definition of measurement error. Classical definitions of error usually center on the notion of *unreplicability*. If a component of a score cannot be replicated by an alternative measure or by the same measure with a time delay, then the component is considered error. But as Cronbach et al. (1972) state, the notion of replicability depends on the context or, as they would say, the type of generalization to be made. Reliability and measurement error then do not reside in the measure, but in the goals of the research. It may be that part of a measure's disturbance is systematic, replicable, and stable. Cook and Campbell (1976) have argued that such components should be called *irrelevancies* and need to be considered as errorful even if they are replicable.

Perhaps the simplest way to refer to the disturbances is that they are unique factors. A unique factor contains both specific and error variance. The important point to remember is that the specific factor is specific relative to the measures included in the study. As an example consider a measure that loads on two factors, A and B, which are uncorrelated. If the study includes only measures that are indicators of A, then the B component of the measure will be considered specific. If only indicators of B are included, then the A component will be specific.

Structural Model

The latent constructs are ordinarily correlated. Perhaps the most serious limitation of factor analysis, as it is customarily practiced, is that the solution is usually constrained to be orthogonal. One may, however, be primarily interested in the correlations between the factors. The traditional factor analysis solution to the analysis of the correlations between factors is to perform a factor analysis on the correlation matrix between factors. The resulting factors are called *second-order* factors. However, since the factors are constructs, albeit unmeasured, any type of structural analysis can be performed on the correlation (covariance) matrix. Models discussed in all the previous chapters are then possible, as well as models discussed in the subsequent chapters. For instance, an example discussed later in this chapter has two latent variables involved in a feedback relationship.

In general the matrix that summarizes the *structural model* has the following form. As in Table 9.1, the rows of the matrix are the n

Table 9.1. Structural Model: The F Variables Are Endogenous and the G Variables Are Purely Exogenous

	G_1	G_2	...	G_m	F_1	F_2	F_3	...	F_n
F_1									
F_2									
F_3									
$\vdots$									
F_n									

constructs that are endogenous and the column includes the complete set of constructs of the measurement model, that is, m exogenous variables and n endogenous variables. Entries into the matrix denote whether a particular construct causes the endogenous variable, and the columns are the causes and the rows the effects. Each row of the matrix then summarizes the structural equation for each unmeasured endogenous construct.

The columns of the matrix are ordered such that the first set of variables includes only the constructs that are purely exogenous. The whole matrix can be divided into two submatrices. The first is a rectangular matrix of endogenous variables by the purely exogenous variables. The second matrix is square, and it is the endogenous variables by the endogenous variables. This partitioning of the matrix is illustrated in Table 9.1. The $n \times m$ matrix of endogenous by exogenous variables represents the causal effects of the exogenous variables.

Now consider $n \times n$ matrix as in Table 9.1. The diagonal values are always set to one and simply represent that a variable is in its own equation. One special type of $n \times n$ matrix is called *lower-triangular.* A lower-triangular matrix contains zero entries above the descending diagonal. Below that diagonal the entries may be nonzero. If the $n \times n$ matrix can be arranged by shifting rows and columns to be lower-triangular, then the structural model is hierarchical; that is, it involves no feedback. If the $n \times n$ matrix cannot be arranged to make it lower-triangular, then the model is nonhierarchical.

Also to be considered are the two final parts of this very general formulation. First, the purely exogenous variables may be correlated and so one must consider the correlation (covariance) matrix for these variables. Second, the disturbances of the n endogenous variables may be correlated with each other or with the purely exogenous variables.

The preceding general formulation can represent every previously discussed model in this text. For instance, the models discussed in Chapters 4 and 6 have a very elementary measurement model. Each measure is a perfect indicator for each construct. Put into more familiar language, each measure is the operational definition for the construct. The matrix for the measurement model would then be a square matrix with ones in the diagonal, that is, an identity matrix. Since each construct is defined by the measure, there are no disturbances for the measurement model. The matrix for the structural model would be lower-triangular for the hierarchical structural models in Chapters 4 and 5 and would not be lower-triangular for the models discussed in Chapter 6.

For the model in Chapter 5 in which an exogenous variable correlates with the disturbance, the measurement model is an identity matrix, and the structural model implies a lower-triangular matrix. For the exogenous variables measured with error in Chapter 5, the errorful measures need a disturbance. The matrix for the measurement model remains an identity matrix if the classical model of measurement error is assumed.

In Chapter 7 there is only a measurement model and its matrix is a column vector (a string of numbers). For the models in Chapter 8 there is a multicolumn measurement model. Generally each row of the matrix has the following restriction: One value is nonzero and the others are zero. Again there is no structural model.

Identification

For the structural model to be identified the correlations (covariances) among the factors must be identified. Assuming that such correlations are identified, the researcher then proceeds to use the rules established in Chapters 4, 5, and 6. For instance, if the researcher can establish that the model among the latent variables is hierarchical and that none of the exogenous variables are correlated with disturbances, multiple regression can be used to estimate the model, assuming the correlations among the latent variables can be estimated. Although estimation is more extensively discussed in the next section, it is important to note here that any of the estimation procedures discussed in previous chapters can be applied to virtually any correlation (covariance) matrix. The correlations can be estimated directly from sample data, estimated indirectly as by factor analysis, or simply postulated by theory. However, if the correlations are indirectly estimated or postulated, one should not employ the standard tests of significance. For instance, it is totally illegitimate to test the correlation between two *unmeasured* variables by the standard t-test of a correlation.

Assuming the correlations among the factors are identified, the status of identification of the structural model is straightforward. Overidentification of the measurement model can in no way aid in the identification of the structural model. Thus, adding more indicators cannot facilitate the estimation of the structural model except in the sense that it may identify the factor correlations. Recall that the identification of the structural model presumes that the factor correlations are identified.

The identification of the measurement model was discussed in the previous chapter. Assuming that loadings and factor correlations are moderate, only two indicators per construct are needed. Three indicators are safer, and four safer still. In some very special cases when the structural model is overidentified, only one indicator of a construct is necessary. Instrumental variable estimation discussed in Chapter 5 is one such an example. When one exogenous variable is known to have a zero path, then such a restriction can buy the identification of a path from another exogenous variable to its indicator. Thus, an overidentified structural model can in some very special cases aid in the identification of the measurement model. More typically the overidentification of the structural model does not help in the identification of the measurement model.

If the model is overidentified overall, it is usually helpful to be able to test for specification error in the various components of the model. One reasonable strategy is to fit a just-identified structural model. Any lack of fit of the model to the correlations can be attributed to specification error in the measurement model. The measurement model can be adjusted to achieve more satisfactory fit, and then the overidentified structural model can be fitted. Any significant increase in lack of fit can then be attributed to specification error in the structural model.

Estimation

Quite clearly the estimation task is quite complex. The measurement model requires a factor analysis type model and the structural model is typically a multiple regression type model. None of the previously discussed methods can handle such a general model. Some special models can be estimated by a two-step procedure. If the structural model is just-identified, the first step is to estimate the measurement model by confirmatory factor analysis. From such an analysis one would obtain the correlations (covariances) among the factors. The second step is to input these correlations to a multiple regression or two-stage least squares program for the estimation of the structural model. As mentioned earlier, one cannot use the standard test of significance for these estimates.

Instead of the two-step procedure, the program LISREL (Jöreskog & Sörbom, 1976) can be used to estimate the parameters of such models by maximum likelihood. The fit of the model can be evaluated by a χ^2 goodness of fit test. With LISREL eight different matrices are set up:

1. The loading matrix for the indicators of the endogenous variables.
2. The loading matrix for the indicators of the exogenous variables.
3. The paths from latent endogenous variables to latent endogenous variables.

4. The paths from latent exogenous variables to latent endogenous variables.
5. The covariance matrix for the latent exogenous variables.
6. The covariance matrix for the disturbances of the latent endogenous variables.
7. The covariance matrix for the disturbances of indicators of the latent endogenous variables.
8. The covariance matrix for the disturbances of indicators of the latent exogenous variables.

In LISREL the variances of the latent endogenous variables cannot be fixed and, therefore, cannot be standardized. Thus, to identify the model, either one loading on each latent endogenous variable must be fixed to some nonzero value, or the variance of the disturbance of the latent variable must be fixed to some nonzero value. The former strategy is to be preferred if results are to be compared across different populations or occasions. LISREL does take the solution and then standardizes the latent variables. This is called the standardized solution in LISREL.

LISREL is not an easy program to learn how to use, nor is it inexpensive to run. Moreover, it is continually undergoing revision and it is in its fourth version. It does, however, provide the most complete solution to the estimation problem of structural models.

EXAMPLES

Three Constructs, Each with Two Indicators

In the previous chapter a model was discussed from Crano et al. (1972). In Figure 9.1 there is added a structural model in which language usage causes work skills which causes arithmetic, each factor having two indicators. Language usage is denoted as L, work skills as W, and arithmetic as A.

In Table 9.2 there are the two matrices that summarize both the measurement and structural models. The first matrix denotes which construct causes each measure. An "X" indicates the measure loads on the factor and "O" denotes it does not. Table 9.2 should not be confused with any variant of the game tic-tac-toe.

The rows of the top matrix in Table 9.2 are the six measures and the columns are the three latent constructs for the model in Figure 9.1. Note that each latent construct causes two different measures. The

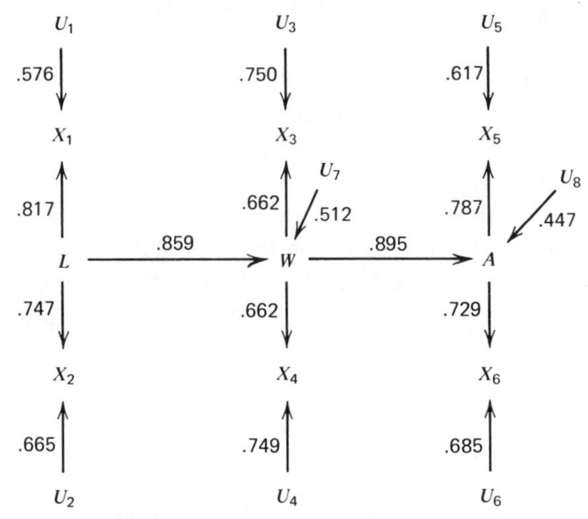

Figure 9.1 Structural model for Crano, Kenny, and Campbell example.

structural model indicated by the bottom matrix in Table 9.2 is hierarchical since the matrix for it is lower-triangular; that is, all the Xs are below the main descending diagonal.

The measurement model can be estimated as in the previous chapter. It is identified since each construct has two indicators. In Chapter 8

Table 9.2. Measurement and Structural Models for Figure 9.1

Measurement Model			
	L	W	A
X_1	X	O	O
X_2	X	O	O
X_3	O	X	O
X_4	O	X	O
X_5	O	O	X
X_6	O	O	X
Structural Model			
	L	W	A
L	O	O	O
W	X	O	O
A	O	X	O

it was noted that the overall fit of the model is good, but there is some problem in the homogeneity between language usage and arithmetic. The measurement model also yields the following estimates of correlations between factors: $\rho_{LW} = .859$, $\rho_{LA} = .768$, and $\rho_{WA} = .895$ (see Figure 8.6). These correlations are the major ingredients in the estimation of the structural model.

Since the structural model is hierarchical, one can use multiple regression to estimate the path coefficients. The path from L to W is simply ρ_{LW} or .859. If paths from both L and W to A are specified, one can solve for them by using the multiple regression formulas 4.5 and 4.6:

$$p_{AL} = \frac{\rho_{AL} - \rho_{AW}\rho_{LW}}{1 - \rho_{LW}^2} = -.003$$

$$p_{AW} = \frac{\rho_{AW} - \rho_{AL}\rho_{LW}}{1 - \rho_{LW}^2} = .898$$

To find the coefficient for the disturbances solve for both

$$(1 - \rho_{LW}^2)^{1/2} = .512$$

$$(1 - \rho_{AW}p_{AW} - \rho_{AL}p_{AL})^{1/2} = .446$$

which are the disturbance paths for W and A, respectively. No special problems arise as a result of having a structural model with unobserved variables, once the correlations (covariances) between the unobserved variables are known. Special considerations do arise if the researcher wishes to estimate a structural model that is overidentified, as in Figure 9.1. Such a model makes the correlations among the latent variables overidentified by the structural model. The program LISREL was developed to estimate these models.

Consider the model in Figure 9.1. It is like the model in Figure 8.6, but in this case there is an overidentified structural model for the latent variables. The LISREL estimates are in Figure 9.1. The $\chi^2(7) = 9.98$ for the model. If a path is added from L to A, $\chi^2(6) = 9.98$. Thus the test of the measurement model is $\chi^2(6) = 9.98$ and the structural model is $\chi^2(1) = .00$. Remember the test of the measurement model is to fit a just-identified structural model. Lack of fit then is due only to specification error in the measurement model. To test an overidentified structural model, one fits that model. Then any increased lack of fit over the just-identified structural model is due to specification error in the structural model.

What is unusual about the structural model in Figure 9.1 is that the path from L to A is virtually zero. Thus the fit of the overidentified model and the just-identified model do not differ and the other parameters are virtually unchanged. Such a result is highly unusual. However, the zero path can be compared against other more standard estimation procedures. If one simply regresses one indicator of A on one indicator of L and one of W, one would obtain on the average

$$p_{AL} = .320 \qquad p_{AW} = .305$$

a result that is radically different from the multiple indicator results. If one unit weights the two indicators of each construct and then regresses the sum of the two indicators of A on the sums for L and W, one would obtain

$$p_{AL} = .298 \qquad p_{AW} = .469$$

At least for this example an unmeasured variable approach yields a conclusion that is very different from multiple regression.

Example with Four Constructs, Two Measured and Two Unmeasured

This example is taken from the Westinghouse Learning Corporation evaluation of Head Start. As some readers no doubt know, there is a great deal of controversy that centers around this work (see Campbell & Erlebacher, 1970). Magidson (1977) has undertaken a reanalysis of these data and the example is taken from his analysis, though the analysis following differs from his in many respects. Magidson's sample consists of 303 white 6-year-old first graders who attended summer Head Start. He chose this group because the original analysis obtained the strongest negative effect for this group. The group consists of 148 children who received Head Start and 155 control children.

For the following analysis six measured variables whose correlations are in Table 9.3 are considered. The variables are mother's education (X_1), father's education (X_2), father's occupation (X_3), income (X_4), and a dummy for Head Start (H), and the outcome variable (Y). The Head Start dummy variable is coded one for those who attended Head Start and zero for those who did not. Note that the correlation between H and Y is negative, indicating that those who completed Head Start had lower test scores.

The proposed causal model is contained in Figure 9.2. For the moment ignore the parameter estimates. The model postulates two

Table 9.3. Head Start Correlations with Residuals above the Main Diagonal[a,b]

X_1	1.000	.000	−.045	.041	.009	.012
X_2	.468	1.000	.032	−.017	.012	−.017
X_3	.241	.285	1.000	.000	−.008	.017
X_4	.297	.209	.407	1.000	.010	−.022
H	−.118	−.084	−.220	−.179	1.000	−.000
Y	.275	.215	.255	.190	−.094	1.000
	X_1	X_2	X_3	X_4	H	Y

[a]Data taken from Magidson (1977).
[b]N = 303.

latent variables. The first is labeled E in Figure 9.2 and it can be viewed as an educational background factor. The second factor relates to socioeconomic outcomes and is labeled S. Obviously the model of latent factors in 9.2 is unrealistic since the measured variables have themselves a structural relationship. For instance, no doubt father's education causes father's occupation. However, it may be plausible to view the child's cognitive skills as a function of the two proposed factors. The educational background factor would directly cause cognitive gain through imitation and the socioeconomic outcome factor

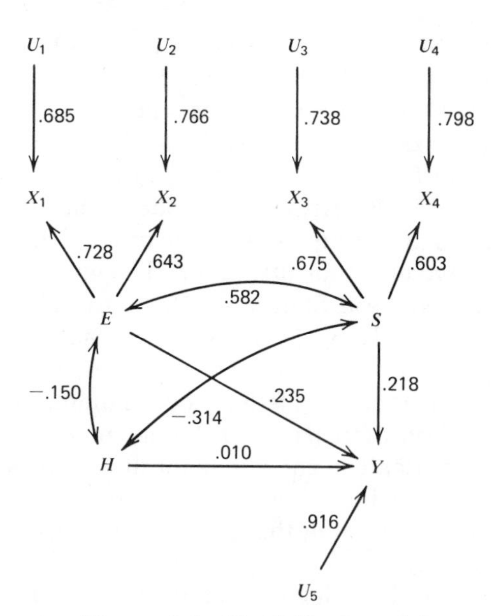

Figure 9.2 Head Start model.

would allow the parents to purchase goods that would increase the child's cognitive skills.

The measurement model for the Head Start example is straightforward. There are four constructs, E, S, H, and Y. E and S each have two indicators, and H and Y have only a single indicator that defines the construct. There is only a single equation for the structural model. The equation is that E, S, and H cause Y. Since the structural model is just-identified, one can use ACOVS to estimate the parameters of the model. Note that the model in Figure 9.2 can be viewed as a simple three-factor model: E, S, and H. The variable Y simply loads on all three factors. The analysis then is rather straightforward: Simply estimate this three-factor model. The first factor, E, has three "indicators" X_1, X_2, and Y; the second, S, also has three "indicators" X_3, X_4, and Y; and the third, H, has two "indicators" Y and H. H's loading on factor H is fixed to one. The solution is presented in Figure 9.2. (Exactly the same estimates were obtained from LISREL and ACOVS.) Not surprisingly there is a high correlation between E and S. Both E and S correlate negatively with H but the socioeconomic outcomes factor correlates more negatively than the educational factor. Just such a difference is to be expected since eligibility into Head Start depended on socioeconomic outcomes. The pattern is reversed for the effects of the factors on cognitive skills. The educational factor has a slightly larger effect than the socioeconomic factor. Head Start has a slightly positive effect which is clearly not statistically significant. However, the effect is not negative as would be obtained had a simple regression analysis been performed with only *measured* X_1, X_2, X_3, and X_4 partialled out. Thus, the structural analysis wipes out the negative effect.

There are a total of 10 free parameters for the Head Start model. Since there are a total of 15 correlations, the number of overidentifying restrictions is five. The $\chi^2(5)$ equals 6.13, which is clearly not significant. In Table 9.3 are the residuals for the model and they are not large. The largest residual occurs in a place where a specification error surely occurs. It is between mother's education (X_1) and father's occupation (X_3) and is negative.

The five overidentifying restrictions of the model are totally due to the measurement model. Recall that the structural model is just-identified. Using the vocabulary of the previous chapter the five overidentifying restrictions can be partitioned as follows:

1. The homogeneity between E and S.
2. The consistency of E for H.
3. The consistency of S for H.

4. The consistency of E for Y.
5. The consistency of S for Y.

As an exercise show that the following 11 vanishing tetrads hold for the model in Figure 9.2 and show that only five are independent:

$$\rho_{13}\,\rho_{24} - \rho_{14}\,\rho_{23} = 0$$

$$\rho_{13}\,\rho_{2H} - \rho_{1H}\,\rho_{23} = 0$$

$$\rho_{14}\,\rho_{2H} - \rho_{1H}\,\rho_{24} = 0$$

$$\rho_{13}\,\rho_{2Y} - \rho_{1Y}\,\rho_{23} = 0$$

$$\rho_{14}\,\rho_{2Y} - \rho_{1Y}\,\rho_{24} = 0$$

$$\rho_{1Y}\,\rho_{2H} - \rho_{1H}\,\rho_{2Y} = 0$$

$$\rho_{13}\,\rho_{4H} - \rho_{1H}\,\rho_{34} = 0$$

$$\rho_{23}\,\rho_{4H} - \rho_{2H}\,\rho_{34} = 0$$

$$\rho_{13}\,\rho_{4Y} - \rho_{1Y}\,\rho_{34} = 0$$

$$\rho_{23}\,\rho_{4Y} - \rho_{2Y}\,\rho_{34} = 0$$

$$\rho_{3Y}\,\rho_{4H} - \rho_{3H}\,\rho_{4Y} = 0$$

Nonhierarchical Relationship between Latent Variables

The preceding models that have been considered have had relatively simple structural models. The model to be considered now contains almost every imaginable complexity. It contains, among other things:

a. Unmeasured endogenous variables with multiple indicators.
b. A nonhierarchical relationship between unmeasured variables.
c. An overidentified structural model.
d. Equality constraints for causal parameters.

The data are taken from Duncan, Haller, and Portes (1971), which has become a classic in the causal modeling literature. The substantive question of their paper concerns itself with the effect of peers on aspirations. The measured variables are parental encouragement of the child (X_1), child's intelligence (X_2), socioeconomic status of the parents (X_3), and the educational and occupational aspirations of the self (Y_1 and Y_2). The same variables were measured for the person named by the child as the best friend (ordered the same as for self: X_4, X_5, X_6, Y_3, and Y_4). There are then a total of 10 variables, 5 for the self and 5 for the friend. The sample consists of 329 adolescent males and the correlations are presented in Table 9.4. In Figure 9.3 there is a model for the data that Duncan, Haller, and Portes postulated with some modifications. First, Duncan et al. postulated that the disturbances in the aspiration measures of the self are correlated. For the sake of simplicity, it is assumed here that such correlations are zero. Second, the model is constrained to be the same for both the self and the friend. Duncan et al. give a compelling reason not to force such equality. They argue that the self named the other child as a friend while the friend did not necessarily make a reciprocal choice. One would then expect that the friend would have more influence on the chooser. However, for illustrative purposes here equality of the coefficients is assumed.

An examination of the model in Figure 9.3 shows that there are six measured exogenous variables, X_1 through X_6. These variables cause two latent endogenous variables, G and H, which Duncan et al. called ambition of self and friend. Each latent construct has two indicators, educational and occupational aspiration. The latent endogenous variables are not caused by all the exogenous variables. The intelligence of the self's friend and the friend's parents' encouragement are assumed not to cause the self's ambition. The two unobserved variables are involved in a feedback relationship. The child's ambition causes his friend's ambition and vice versa.

The structural equations are as follows for the measurement model:

$$Y_1 = eG + iE_1$$

$$Y_2 = fG + jE_2$$

$$Y_3 = eH + iE_3$$

$$Y_4 = fH + jE_4$$

and for the structural model

$$G = aX_1 + bX_2 + cX_3 + dX_6 + gH + hU_1$$

$$H = aX_4 + bX_5 + cX_6 + dX_3 + gG + hU_2$$

In Table 9.5 is the measurement model in matrix form. Note that since the exogenous variables are measured, they each define their own factor and have loadings of one. Both G and H have two indicators. In Table 9.6 there is the structural model. The purely exogenous variables

Table 9.4. Peer Aspiration Example with Residuals of the Most General Model above the Diagonal[a,b]

X_1	1.000	.000	.000	.000	.000	.000	−.025	.019	−.012	.009
X_2	.184	1.000	.000	.000	.000	.000	.018	−.014	−.006	.004
X_3	.049	.222	1.000	.000	.000	.000	−.033	.025	−.003	.002
X_4	.115	.102	.093	1.000	.000	.000	.006	−.005	−.036	.025
X_5	.078	.336	.230	.209	1.000	.000	.018	−.014	.011	−.008
X_6	.019	.186	.271	−.044	.295	1.000	.039	−.030	−.012	.009
Y_1	.214	.411	.324	.076	.300	.293	1.000	.000	.092	−.024
Y_2	.274	.404	.405	.070	.286	.241	.625	1.000	−.028	−.011
Y_3	.084	.260	.279	.199	.501	.361	.422	.327	1.000	.000
Y_4	.112	.290	.305	.278	.519	.411	.327	.367	.640	1.000
	X_1	X_2	X_3	X_4	X_5	X_6	Y_1	Y_2	Y_3	Y_4

[a]Data taken from Duncan, Haller, & Portes (1971).
[b]N = 329.

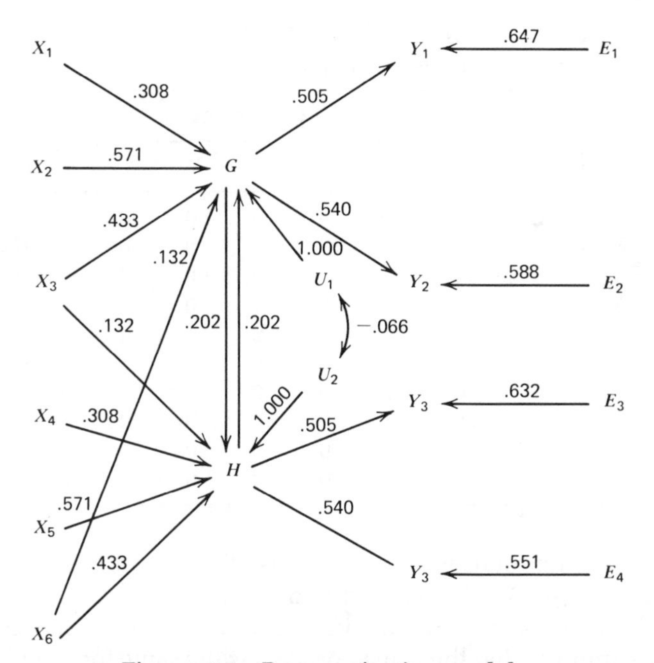

Figure 9.3 Peer aspiration model.

are X_1 through X_6, and G and H are endogenous. The model is not hierarchical since G causes H and vice versa.

LISREL must be used to estimate the parameters for the model in Figure 9.3 since the structural model is overidentified. Equality constraints in LISREL can only be placed on the unstandardized model. Recall that in LISREL the latent endogenous variables are not standardized. Any equality constraint does not necessarily hold for the standardized solution. It was, therefore, decided to fix the disturbance path h to one, and the solution is presented in Figure 9.3 without the correlations among the purely exogenous variables.

There are a total of 17 free parameters and 45 correlations. The 17 free parameters are

a. Nine correlations among the exogenous variables.

b. Four paths from the exogenous variables to the latent endogenous variables.

c. One path from the endogenous variable to the other.

d. The correlation between the disturbances of G and H.

e. Two paths from G and H to their two indicators.

Table 9.5. Measurement Model for the Duncan, Haller, and Portes Example

	Constructs							
Measures	X_1	X_2	X_3	X_4	X_5	X_6	G	H
X_1	1	0	0	0	0	0	0	0
X_2	0	1	0	0	0	0	0	0
X_3	0	0	1	0	0	0	0	0
X_4	0	0	0	1	0	0	0	0
X_5	0	0	0	0	1	0	0	0
X_6	0	0	0	0	0	1	0	0
Y_1	0	0	0	0	0	0	e	0
Y_2	0	0	0	0	0	0	f	0
Y_3	0	0	0	0	0	0	0	e
Y_4	0	0	0	0	0	0	0	f

There are then a total of 28 restrictions. The test of these restrictions yields a $\chi^2(28) = 35.00$ which is nonsignificant. These 28 restrictions can be divided in the following way. First, 13 of the restrictions are due to the assumption that the causal process is the same for friend and self. There are 13 equality constraints because a, b, c, d, g, e, and f are duplicated and the following correlations among the exogenous variables are assumed to be equal: $\rho_{15} = \rho_{24}$, $\rho_{16} = \rho_{34}$, $\rho_{26} = \rho_{35}$, $\rho_{12} = \rho_{45}$, $\rho_{13} = \rho_{46}$, and $\rho_{23} = \rho_{56}$. Thus the equality constraints take care of 13 of the 28 overidentifying restrictions. There are 15 remaining constraints on the measurement and structural model. There are two constraints on the structural model since there are two zero paths from the exogenous variables to each endogenous variable. There is then one extra instrument for the equations for both G and H, creating two overidentifying restrictions. The remaining 13 restrictions are due to the measurement model. One of the 13 constraints is homogeneity between G and H, and

Table 9.6. Structural Model for the Duncan, Haller, and Portes Example

	Exogenous Variables							
Endogenous Variables	X_1	X_2	X_3	X_4	X_5	X_6	G	H
G	a	b	c	0	0	d	0	g
H	0	0	d	a	b	c	g	0

the remaining constraints are the consistency of the epistemic correlations of both G and H for X_1 through X_6.

In Table 9.7 the overall χ^2 has been partitioned into the three different types of restrictions. The three possible types of specification error are equality of parameters, the structural model, and the measurement model. To obtain the χ^2 to test the equality constraints, first a model is estimated in which the equality constraints are not forced to hold. The resulting $\chi^2(15)$ for this model equals 26.83 and subtracting this from the overall χ^2 yields a test of the equality constraints. To test the constraint of the structural model, paths are added from friend's intelligence to self's ambition and from self's intelligence to friend's ambition with no equality constraints. The χ^2 equals 25.21. Subtracting this χ^2 from the previous χ^2 yields $\chi^2(2) = 1.62$ which tests whether the path from X_2 to H and X_5 to G are zero given that the paths from X_1 to H and from X_4 to G are zero. (This turns out to be equivalent to testing whether the paths from X_4 to G and X_1 to H are zero given zero paths from X_2 to H and X_5 to G.) The χ^2 that remains then tests the measurement model. There does appear to be specification error in the measurement model. An examination of the residuals shows a large value for $r_{Y_1Y_3}$. The model should be respecified to allow for correlated disturbances.

The reader can now appreciate the power of LISREL to estimate structural models. If a model of the complexity of Figure 9.3 can be estimated, then no model should pose any problem. The researcher still must carefully study the model and determine the status of identification of the measurement and structural model.

Causal Chain, Single-Indicator Model

Perhaps one of the most interesting models is the causal chain with each construct having only a single indicator. As shall be seen, some of the parameters are overidentified while others are underidentified.

Table 9.7. Partition of χ^2 into Three Components of Specification Error

	χ^2	*df*
Equality constraints	8.17	13
Structural model	1.62	2
Measurement model	25.21	13
Total	35.00	28

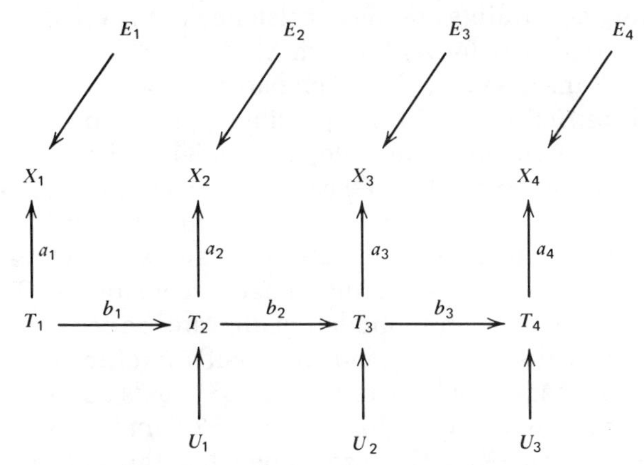

Figure 9.4 First-order autoregressive model with measurement error.

In Figure 9.4 there is a model of a simple chain. As is typical of such models the chain is composed of a single variable measured at multiple points in time. In this case the variable is the number of citations of 240 chemists. The citations were counted by Hargens, Reskin, and Allison (1976) for four years, 1965 through 1968. In Table 9.8 are the correlations for the four waves of data. In Figure 9.4, X denotes the measured number of citations and T the "true" number of citations. Note T is caused only by the prior value of T. The technical name for such a process is *first-order autoregressive*. The correlational structure generated by a first-order autoregressive process is called a *simplex* (Humphreys, 1960). A simplex tends to have lower and lower correlations as one moves from the main descending diagonal. The mathematical

Table 9.8. Correlation Matrix for the Chemist Example: Correlations below the Main Diagonal, Variances on the Diagonal, and Residuals to the Model with Equal Reliabilities above the Diagonal[a,b]

X_1	2.590	.002	−.014	.018
X_2	.515	2.150	.005	−.006
X_3	.469	.613	2.730	−.002
X_4	.436	.520	.552	2.510
	X_1	X_2	X_3	X_4

[a]Data taken from Hargens, Reskin, and Allison (1976).
[b]N = 240.

definition of a simplex matrix is that the variables can be ordered such that

$$\rho_{X_iX_k} = \rho_{X_iX_j}\,\rho_{X_jX_k}$$

$$i < j < k$$

Stated in words, if one partials any variable that comes between two variables in the causal chain, the partial correlation is zero.

Only the true score is governed by a first-order autoregressive process and only its correlational structure is a simplex. The measured score cannot be described so simply. Actually the measured score is assumed to be the sum of two autoregressive processes: the true score and the errors of measurement. The errors are a degenerate case of a first-order autoregressive process since the regression of E on a prior value of E is zero. The correlational structure of the measured scores is called a *quasi-simplex*; that is, it would be a simplex if the scores were corrected for attenuation due to measurement error. A quasi-simplex differs from a pure simplex in that $\rho_{X_iX_k \cdot X_j}$ does not equal zero and is usually positive.

Estimation in the quasi-simplex concerns two types of parameters. First, there are the paths from each construct to its measure that constitutes the measurement model. Second there is the estimation of the causal chain or the structural model. It has long been known that variants of principal components analysis do not even come close to estimating the parameters of either a simplex or its cousin the quasi-simplex. Diagonal factoring does estimate parameters of a simplex model, but requires prior knowledge of communalities to estimate quasi-simplex parameters. The approach here is first to discuss path analytic solutions following Heise (1969) and Wiley and Wiley (1970) and, second, to elaborate a confirmatory factor analysis approach to estimation (Jöreskog, 1970; Werts, Jöreskog, & Linn, 1971).

For the model in Figure 9.4 there are six correlations

$$\rho_{12} = a_1a_2b_1 \qquad \rho_{23} = a_2a_3b_2$$

$$\rho_{13} = a_1a_3b_1b_2 \qquad \rho_{24} = a_2a_4b_2b_3$$

$$\rho_{14} = a_1a_4b_1b_2b_3 \qquad \rho_{34} = a_3a_4b_3$$

Note that there are seven parameters and six correlations, and thus the model must be underidentified.

Even though the model is underidentified, the parameter estimates for b_2^2, a_2^2, and a_3^2 are overidentified.

$$a_2^2 = \frac{\rho_{12}\rho_{23}}{\rho_{13}} = \frac{\rho_{12}\rho_{24}}{\rho_{14}}$$

$$a_3^2 = \frac{\rho_{13}\rho_{34}}{\rho_{14}} = \frac{\rho_{23}\rho_{34}}{\rho_{24}}$$

$$b_2^2 = \frac{\rho_{14}\rho_{23}}{\rho_{12}\rho_{34}} = \frac{\rho_{13}\rho_{24}}{\rho_{12}\rho_{34}}$$

Both a_2^2 and a_3^2 are reliabilities and the positive root is ordinarily taken. The sign of b_2 is the same as the sign of ρ_{23}. Since the preceding parameters are overidentified, one can solve for an overidentifying restriction. It is

$$\rho_{14}\rho_{23} - \rho_{13}\rho_{24} = 0$$

This is a vanishing tetrad which can be tested by checking whether the second canonical correlation is zero. The result of such a test yields $\chi^2(1) = .508$ which is not significant, indicating that the overidentifying restriction is met.

The estimates of a_2, a_3, and b_2 for the chemists' data are

a_2: .820, .784

a_3: .807, .771

b_2: .970, .926

The two estimates of each parameter are roughly similar, which is not surprising since the overidentifying restriction holds. The "low" reliabilities are discussed later.

One is not able to estimate a_1, a_4, b_1, and b_3. One can, however, solve for the product terms, a_1b_1 and a_4b_3, if a_2, a_3, and b_2 are known. They can be estimated by

$$a_1b_1 = \frac{\rho_{12}}{a_2} = \frac{\rho_{13}}{a_3b_2}$$

$$a_4b_3 = \frac{\rho_{34}}{a_3} = \frac{\rho_{24}}{a_2b_2}$$

Since all the parameters of the model are also correlations, they should be within $+1$ and -1. However, given sampling error, an estimate may fall out of this range. Note that if the reliabilities or autoregressive parameters are zero or near zero the model is empirically underidentified.

In general given k waves of measurement, the reliabilities of the first and last wave are not identified. Also not identified is the path from the true score at the first wave to the true score at the second wave and the path from the true score at the next to the last wave to the true score at the last wave. All the other parameters are identified if k equals three, and overidentified if k is greater than three. All the overidentifying restrictions are of vanishing tetrad form:

$$\rho_{im}\rho_{jk} - \rho_{ik}\rho_{jm} = 0$$
$$i < j < k < m$$

Since the model is partially underidentified, various strategies have been put forward to bring about identification. Heise (1969) suggested that one might assume that $a_1 = a_2 = a_3 = a_4$. This implies that reliability does not change over time. This yields a just-identified model for the three-wave case and an additional overidentifying restriction for the four-wave case. Alternatively one might force what seem to be very ad hoc specifications of $a_1 = a_2$ and $a_3 = a_4$ or even more implausibly $a_1 = a_3$ and $a_2 = a_4$.

Wiley and Wiley (1970) have argued that it may be more plausible to assume that the error variance, $V(E)$, and not the reliability, is constant over time. Unlike the previous models this requires the use of the unstandardized metric and refers to the covariance matrix, not the correlation matrix. The measurement model is

$$X_i = T_i + E_i$$

where $V(E_1) = V(E_2) = V(E_3) = V(E_4)$. The measures X are said to be tau equivalent since their error variances are assumed to be equal. The structural model is

$$T_{i+1} = b_i T_i + V_i$$

The covariance matrix for the four-wave case is in Table 9.9. One can

Table 9.9. Theoretical Covariance Matrix with Tau Equivalent Tests and a First-Order Autoregressive Process

X_1	$V(T_1) + V(E)$			
X_2	$b_1V(T_1)$	$V(T_2) + V(E)$		
X_3	$b_1b_2V(T_1)$	$b_2V(T_2)$	$V(T_3) + V(E)$	
X_4	$b_1b_2b_3V(T_1)$	$b_2b_3V(T_2)$	$b_3V(T_3)$	$V(T_4) + V(E)$
	X_1	X_2	X_3	X_4

then estimate $V(E)$ by both

$$V(X_2) - \frac{C(X_1,X_2)C(X_2,X_3)}{C(X_1,X_3)}$$

and

$$V(X_3) - \frac{C(X_2,X_3)C(X_3,X_4)}{C(X_2,X_4)}$$

The estimates of $V(T_i)$ can now be computed by $V(X_i) - V(E)$. One can then proceed to estimate b_1, b_2, b_3, b_1b_2, b_2b_3, and $b_1b_2b_3$. Note that what is analyzed is the *covariance* matrix not the correlation matrix.

The third solution to this underidentification problem has not been previously suggested in this context. It is, however, very popular in the time series literature. It assumes that the autoregressive process is stationary, that is, that $b_1 = b_2 = b_3$. Such an assumption brings about identification for the four-wave case but adds no further overidentifying restrictions. It does not bring about identification for the three-wave case as do assumptions about constant reliabilities or error variances.

All the preceding strategies have the particular problem that there is no clear way to pool the overidentified information. LISREL can be employed to estimate all the preceding models. Inputting the correlation matrix into LISREL and forcing the equality of the paths from T_t to T_{t+1} does not result in equal reliabilities since one cannot standardize in LISREL T_2, T_3, or T_4. To estimate a model with equal reliabilities then one forces the disturbance paths to the measures to be equal. In Table 9.10 are the estimates for the four models. All the estimates are standardized except the error variances and the estimates for the equal error variance model. For the first model there are no constraints; this makes a_1, a_4, b_1, b_3, $V(E_1)$, and $V(E_4)$ underidentified. The fit of the model is quite good. Assuming equal reliabilities does not result in a significant increase in lack of fit. Note that either due to rounding error or small specification error, the a parameters are not exactly equal.

The assumption of equal stabilities, $b_1 = b_2 = b_3$, yields the result that the reliabilities of the first and last waves are lower than the two middle waves. Finally, the model with equal error variances also fits, although not quite as well as the model with equal reliabilities. Table 9.10 gives the unstandardized parameter estimates for this last model.

The reader is still probably puzzled by the low autocorrelation for true citations. Understanding the implicit assumptions of the autoregressive model can explain the low correlation. A test–retest correlation can be less than one for three very distinct reasons: (a) true change in causes, (b) measurement error, and (c) change in structure. The first two

Table 9.10. Model Parameter Estimates for Autoregressive Models; All Parameters Are Standardized Except Error Variances and the Equal Error Variance Model

	Free	Equal Reliabilities	Equal Stabilities	Equal Error Variance
a_1	—	.802	.675	1.000
a_2	.807	.803	.807	1.000
a_3	.797	.800	.797	1.000
a_4	—	.802	.732	1.000
b_1	—	.796	.945	.680
b_2	.945	.946	.945	1.120
b_3	—	.864	.945	.803
$V(E_1)$	—	.925	1.410	.857
$V(E_2)$	.748	.768	.748	.857
$V(E_3)$	.996	.975	.996	.857
$V(E_4)$	—	.896	1.165	.857
χ^2	.549	.587	.549	1.553
df	1	2	1	2

reasons are most certainly already known by the reader, while the third may not be. Assume that X is measured at two distinct points in time, say 1 and 2. The variable X_1 is defined as the sum of its causes, $\Sigma a_i Z_{i1}$ and X_2 similarly, $\Sigma b_i Z_{i2}$. Assume also that none of the causes change and they are all orthogonal. The test-retest correlation will only be one if $a_1/b_1 = a_2/b_2 = a_3/b_3 = \cdots = a_n/b_n$. Thus if the causal structure changes over time, the test–retest correlation is not unity. There is then the implicit assumption that the *structural equations* do not change over time for the data to be first-order autoregressive. In Chapter 12 this assumption is called quasi-stationarity and is discussed in much more detail.

A second implicit assumption of an autoregressive process is homogeneous stability: All true causes change at the same rate; that is, they all have the same autocorrelation. If a measure is multidimensional and the different dimensions change at different rates, the process cannot ordinarily be described as first-order autoregressive. Moreover, *all* the causes that are totally unstable over time are defined as errors of measurement. Thus the reliability estimates that were obtained for the citations of chemists are not low. It is true that if someone else did the counting this new count should correlate very highly with the old count. Thus it is not the reliability of the counts *per se* but the reliability of their ability to indicate some latent trait. The actual number of citations is not a perfectly valid indicator of "citability." See Hargens, Reskin, and Allison (1976) for an alternative explanation.

Composite Cause

In Figure 9.5 is a model that is very different from all others that have been considered in this text. The variable S is unmeasured and has no disturbance. It is assumed to be a composite or linear function of X_1, X_2, X_3, and X_4. The variable S is called a composite cause because a composite of X_1 through X_4 causes Y_1, Y_2, and Y_3. Note that S mediates the causal effect of the X variables on the Y set. Composite variables are more commonly called indices. For instance, one could index a child's aggressiveness by combining physical and verbal aggression. Population change is indexed by number of births minus the number of deaths plus the number of immigrants minus the number of emigrants.

The reader should compare the role of S in Figure 9.5 with the role of say L in Figure 9.1. They are both unmeasured, but once the coefficients leading into S are determined, S is exactly known. However, L can be estimated only with imperfect reliability. The X variables cause S in Figure 9.5, whereas in Figure 9.1 they are caused by L. Finally it is assumed that S's X variables are errorfree, while the indicators of L are unreliable. Clearly there is only a superficial similarity between S and L. It is shown later that there are differences in estimating the two models also.

The four components of S are father's education, mother's education, father's occupation, and average parental income making S a

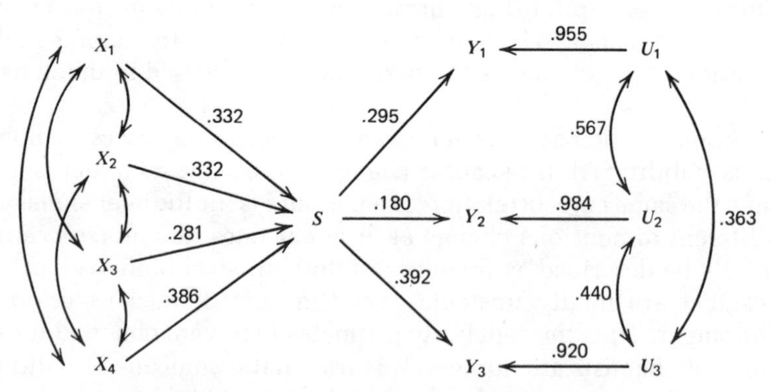

Figure 9.5 Composite cause model.

socioeconomic composite. The Y variables are as follows: mental ability, high school grades, and college plans. Data are taken from Hauser (1973), and the correlations are reproduced in Table 9.11.

To estimate the parameters of the model, first set up the following canonical correlation analysis. The X variables form one set and the Y variables form the second set. The paths from the X variables to S are given by the canonical variate coefficients for the X variables. The estimate of S obtained from such a canonical analysis is

$$.332X_1 + .332X_2 + .281X_3 + .386X_4 \qquad [9.1]$$

To determine the path from S to Y_i note that it equals r_{SY_i}. Thus for the path from S to Y_1 one computes the covariance between Y_1 and Equation 9.1 which is

$$\begin{aligned} p_{Y_1S} &= r_{SY_1} \\ &= p_{X_1S}r_{X_1Y_1} + p_{X_2S}r_{X_2Y_1} + p_{X_3S}r_{X_3Y_1} + p_{X_4S}r_{X_4Y_1} \\ &= (.322)(.244) + (.322)(.230) + (.281)(.212) + (.386)(.203) \\ &= .295 \end{aligned}$$

The paths from S to Y_2 and Y_3 are similarly defined:

$$\begin{aligned} p_{Y_2S} &= (.332)(.151) + (.332)(.149) + (.281)(.127) + (.386)(.116) \\ &= .180 \\ p_{Y_3S} &= (.332)(.306) + (.332)(.269) + (.281)(.299) + (.386)(.304) \\ &= .392 \end{aligned}$$

Table 9.11. Correlations for Socioeconomic Variables and Outcomes[a,b]

X_1	1.000						
X_2	.505	1.000					
X_3	.494	.318	1.000				
X_4	.389	.291	.523	1.000			
Y_1	.244	.230	.212	.203	1.000		
Y_2	.151	.149	.127	.116	.586	1.000	
Y_3	.306	.269	.299	.304	.435	.469	1.000
	X_1	X_2	X_3	X_4	Y_1	Y_2	Y_3

[a]Data taken from Hauser (1973).
[b]N = 3427.

The correlations among disturbances are simply partial correlations: $r_{Y_1Y_2 \cdot S}$, $r_{Y_1Y_3 \cdot S}$, and $r_{Y_2Y_3 \cdot S}$.

The model is overidentified and the set of overidentifying restrictions are of the form

$$\rho_{X_iY_k}\rho_{X_jY_m} - \rho_{X_iY_m}\rho_{X_jY_k} = 0$$

Note that this set of restrictions is equivalent to assuming homogeneity between the X and Y set (see the previous chapter), which is why the second canonical correlation can be used to test for homogeneity between constructs. The degrees of freedom are $(n - 1)(m - 1)$ where n is the number of variables in set X and m the number in set Y. The $\chi^2(6) =$ 11.31, which is not significant at the .05 level. The program LISREL produced the same parameter estimates and $\chi^2(6) = 11.33$.

Normally models for composite causes are much more complicated. The reader should consult Hauser (1973) and Hauser and Goldberger (1971) for more details. The reader should note that while canonical correlation analysis is useful in estimating causal models, the canonical correlation itself does not estimate a meaningful structural parameter.

CONCLUSION

Five specific examples have been analyzed. Quite clearly, structural modeling with unobserved variables is possible. As with any model one must take care in the specification and identification of such models. Always check the estimates and residuals to determine if the results are meaningful. To some extent the measurement model can be separately considered. First, the measurement model can be specified and, if it is identified, one can proceed to estimate the covariances among the latent variables. Then one can specify the structural model and, if it is identified, go ahead and estimate it. Occasionally an overidentified structural model can bring about the identification of a previously underidentified measurement model.

The fit of the model can be tested as follows. First the measurement model can be estimated with a just-identified structural model. Any significant lack of fit can be attributed to the measurement model. Then the full model can be estimated and the increased lack of fit can be attributed to the structural model.

LISREL is a powerful tool for estimating models with latent variables. When working with the program, one should take care not to fall prey to the many common traps:

1. Check to make sure one has not obtained a local minimum by altering the starting values.
2. Check the degrees of freedom to see if it agrees with a simple counting.
3. Do not rely on the program to determine if the model is identified. Try to independently establish the status of identification.
4. Carefully study the parameter estimates and look for anomalous results.

When I run LISREL I presume I have made an error. I check and recheck my results. As George Harrison put it, "with every mistake, we must surely be learning."

10

Causal Model and True Experiments

Although the focus of this text is the analysis of nonexperimental data, in this chapter the topic is true experimentation. Structural modeling can add new insights to true experiments as well as encourage researchers to apply nonexperimental methods even within experimental settings. The first section of the chapter discusses the logic of experimentation. The second section considers alternative dummy coding schemes. The third section presents an example that uses path analytic techniques to estimate structural parameters. The fourth section discusses the advantages of a multiple regression approach to experimental data. The fifth section considers alternative structural formulations of repeated measures designs.

LOGIC OF EXPERIMENTATION

The key feature of the true experiment is *random assignment of experimental units to conditions*. To randomly assign units the experimenter must be able to control or manipulate the causal variable. Occasionally natural experiments like the draft lottery occur (Staw, 1974), but in the main randomization requires that the experimenter control the causal variable. This control has brought about the term that is used to refer to the exogenous variable in an experiment: the *independent variable*. This variable is called independent since it is under experimental control and is not related to any other important exogenous variables. The endogenous variable is called the dependent variable since any covariation between it and the independent variable is brought about through the independent variable.

The requirement of random assignment forces social scientists to choose as independent variables those variables that are under experimental control. To some degree experimentation has locked social science and particularly psychology into an epistemology that outside forces determine the behavior of the individual. Since stimulus variables can ordinarily be manipulated whereas response variables can only be manipulated with difficulty, it is hardly surprising that experimentalists are predisposed to a stimulus–response approach to human behavior (see Chapter 12).

The experimental researcher possesses a vast arsenal of experimental designs from which to choose. For the moment consider experimental designs commonly called between-subject designs. Persons are randomly assigned to receive one particular treatment. In within-subject designs persons receive multiple treatments.

Imagine a two-group experiment in which Np subjects are in the experimental group and $N(1 - p)$ or Nq are in the control group, N being the total number of persons and p the probability of being assigned to the experimental group. Let X be a dummy variable such that when the subject is in the experimental group he or she receives one and if in the control group, the person receives a zero. Calling the dependent variable Y, it follows that $V(X) = pq$ and $C(X,Y) = (\bar{Y}_E - \bar{Y}_C)pq$ where $\bar{Y}_E$ is the experimental group mean and $\bar{Y}_c$ the control group mean. The regression of Y on X is therefore

$$\begin{aligned} b_{YX} &= \frac{C(X,Y)}{V(X)} \\ &= \frac{(\bar{Y}_E - \bar{Y}_C)pq}{pq} \\ &= (\bar{Y}_E - \bar{Y}_C) \end{aligned}$$

that is, the mean difference between the experimental and control group. Not surprisingly, going from 0 to 1 in X, the dependent variable increases by $(\bar{Y}_E - \bar{Y}_C)$.

In Figure 10.1 there is a path diagram. In this case X is the dummy variable, 1 for treatment and 0 for control, and Y is the dependent variable and U is the residual from the regression of Y on X or $Y - b_{YX}X$. The important point about experimentation is that one *knows* that in the population the disturbance is uncorrelated with the exogenous variable. Because of random assignment the independent variable

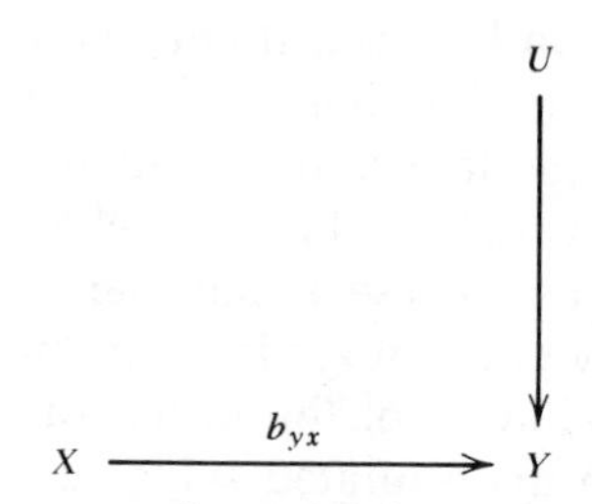

Figure 10.1 Path model for the experiment.

must be uncorrelated with all other causes of the dependent variable. However, in any sample the disturbance is correlated with an exogenous variable due to sampling error, making the guarantee of zero correlation good only in the long run. Clearly, experimentation provides the researcher a specification of great importance. In Chapter 4 it was shown that the key assumption for the causal interpretation of regression coefficients is that the disturbance is uncorrelated with the exogenous variables; however, in experimentation this is guaranteed!

Of course, demonstrating a causal effect is only part of the game. One must attach to that effect a verbal interpretation. The result must be put within some theoretical context. Moreover, often one must rule out the hypothesis that the independent variable is *confounded* with some irrelevant variable. Our labeling the independent variable something in particular does not insure that it is in fact that variable; Cook and Campbell (1976) call this problem *construct validity*. Experimentation in and of itself rules out a number of classic confounding variables like history, maturation, and selection, but numerous other artifacts like experimenter bias and demand characteristics remain. However, the strength of causal inference within true experiments gives it the highest degree of *internal validity* (Campbell & Stanley, 1963) of any method discussed in this text.

Another virtue of true experiments is the possibility of uncorrelated exogenous variables. This strength of experiments rests again on the fact that the experimenter controls the exogenous variable. Consider a 2 $\times$ 2 design. One independent variable, called A, has two levels, A_1 and A_2, and the other, called B, also has two levels, B_1 and B_2. There are then four cells in the design, A_1B_1, A_1B_2, A_2B_1, and A_2B_2. With multiple regression three dummy variables are created, that is, as many dummies as there are cells in the design less one. In Table 10.1 there is what is called the *model matrix* or *design matrix* (Bock, 1975). The columns are the cells of the design and the rows are the dummy variables. The first dummy, X_1, assigns a 1 to subjects in A_1 and a -1 to those in A_2.

Table 10.1. Model Matrix for a 2 × 2 ($A \times B$) Factorial Design

Dummy Variable	Cell			
	A_1B_1	A_1B_2	A_2B_1	A_2B_2
X_1:A	1	1	−1	−1
X_2:B	1	−1	1	−1
X_3:$A \times B$	1	−1	−1	1

Obviously this dummy tests the A main effect. The second dummy, X_2, assigns a 1 to those in the B_1 cells and a -1 to those in the two B_2 cells and is a test of the B main effect. The last dummy, X_3, is formed by taking a product of the main effect dummies, X_1X_2, and tests interaction.

The textbook recommendation for factorial design is to have equal number of subjects in each cell, say n. Examine now the correlation between X_1 and X_2. First, all three means are zero since half the subjects have a score of 1 and the other half have -1. Figuring ΣX_1X_2, it is n for A_1B_1, $-n$ for A_1B_2, $-n$ for A_2B_1, and n for A_2B_2 all of which sums to zero. Working through ΣX_1X_3 and ΣX_2X_3 again zero is obtained. Thus, in this case and in general, given factorial design and equal number of subjects, the main effects and their interactions are all unconfounded. Multicollinearity evaporates as a problem. Actually the whole strategy of *analysis of variance* is based on this fact. Given factorial design, the multiple correlation squared (the between group sums of squares) can be partitioned unambiguously into separate effects. The analysis of variance can be viewed as a computational shortcut for balanced (equal n) factorial designs.

DUMMY VARIABLES

Consider again the dummy variables and model matrix. The values of the model matrix may be designated λ_{ij} where the first subscript i refers to the dummy variable X_i and the second subscript j refers to the column which in turn refers to a cell of the factorial design. The mean of a dummy variable X_i is

$$\bar{X}_i = \frac{\sum_j n_j \lambda_{ij}}{N} \qquad [10.1]$$

where n_j is the number of persons in the jth cell and N is the total number of subjects. Since many coding schemes require $\sum_j \lambda_{ij} = 0$, it is useful to note that $\sum_j \lambda_{ij} = 0$ will always imply that $\overline{X}_i$ is zero when n_j is constant across all conditions. The covariance between two dummies is given by

$$C(X_i, X_k) = \frac{\sum_j n_j \lambda_{ij} \lambda_{kj}}{N} - \frac{\sum_j n_j \lambda_{ij} \sum_j n_j \lambda_{kj}}{N^2} \qquad [10.2]$$

If the covariance is zero, the two dummies are said to be orthogonal. If $\sum_j \lambda_{ij} = \sum_j \lambda_{kj} = \sum_j \lambda_{ij}\lambda_{kj} = 0$, then 10.2 will also be zero if the cell sizes are equal. It is noted here that the variance of a dummy is a special case of Equation 10.2

$$C(X_i, X_i) = \frac{\sum n_j \lambda_{ij}^2}{N} - \left(\frac{\sum n_j \lambda_{ij}}{N}\right)^2$$

A researcher is presented a choice in the coding of multilevel independent variables. As an example consider a four-level independent variable and three different types of coding: Helmert, deviation, and effects coding.

Table 10.2. Three Different Methods of Coding a Multilevel Nominal Variable

		Level			
		I	II	III	IV
	X_1	1	−1/3	−1/3	−1/3
Helmert:	X_2	0	1	−1/2	−1/2
	X_3	0	0	1	−1
	X_1	1	−1/3	−1/3	−1/3
Deviation:	X_2	−1/3	1	−1/3	−1/3
	X_3	−1/3	−1/3	1	−1/3
	X_1	1	0	0	−1
Effects (simple):	X_2	0	1	0	−1
	X_3	0	0	1	−1

In Table 10.2 there are examples of these three types of codes. One use for Helmert codes would be in the case of hierarchically ordered groups. For instance, imagine an educational evaluation that involves a considerable amount of training of students. An experiment was designed with four groups:

1. Pure control.
2. Training only.
3. Training plus one month's use.
4. Training plus six months' use.

Helmert contrasts could be used to contrast group 1 against groups 2, 3, and 4; group 2 against groups 3 and 4; and group 3 against group 4. The first contrast would test the combined effect of the innovation, the second would test whether use was of value, and the last contrast would test whether 6 months' use is better than one month's. Note that if there is equal *n* in the groups, all the Helmert contrasts have zero mean and are orthogonal.

Deviation coding, the second method in Table 10.2, seems to compare a group with the other groups. In the univariate case if level 1 is coded with a 1 and the other three levels with $-1/3$, the univariate regression of the dependent variable on X_1 yields

$$\bar{Y}_{\mathrm{I}} - \frac{\bar{Y}_{\mathrm{I}} + \bar{Y}_{\mathrm{II}} + \bar{Y}_{\mathrm{III}} + \bar{Y}_{\mathrm{IV}}}{4}$$

However, if the other deviation codes are created as in Table 10.2, the partial regression of the dependent variable on X_1 partialling out X_2 and X_3 yields

$$\frac{\bar{Y}_{\mathrm{I}} - \bar{Y}_{\mathrm{IV}}}{2}$$

Although a deviation coding seems to compare one group with the other groups, in the multivariate case it actually compares one group with one other group. To see this, note that X_1, X_2, and X_3 all compare groups 1, 2, and 3. The only part that is unique about each contrast is the comparison of each group with the fourth group. Thus, when the other variables are partialled out, only the comparison of each group with group four remains. Deviation codings are clearly not orthogonal, and are in fact always negatively correlated.

The final sets of codes are called effects coding or simple coding by Bock (1975). These codes work in just the opposite way as the deviation codes. In the univariate case for the regression coefficient X_1 yields

$$\frac{\bar{Y}_{\mathrm{I}} - \bar{Y}_{\mathrm{IV}}}{2}$$

and the multivariate case

$$\overline{Y}_{\rm I} - \frac{\overline{Y}_{\rm I} + \overline{Y}_{\rm II} + \overline{Y}_{\rm III} + \overline{Y}_{\rm IV}}{4}$$

They are called effects codes because they yield regression coefficients identical to analysis of variance effect estimates. Effects coding is not orthogonal even in the equal n case and usually yields positive correlations. For example, with equal n and any number of categories, the correlation between two dummies is 1/2.

Clearly the creation of dummies must be done with care. They do not always estimate or test what they seem to be. There are two helpful rules that can aid one in setting up dummy variables. However, the rules should not be slavishly followed if they conflict with theoretical considerations. First try to keep the dummies as orthogonal as possible. A second helpful rule is to express dummies in mean deviation form, which yields three advantages:

1. It can be quickly determined if two dummies are orthogonal by $\Sigma n_j \lambda_{ij} \lambda_{kj}$.
2. The intercept in the regression equation estimates the grand mean.
3. If interaction contrasts are formed by the multiplication of main effect dummies, the interaction is ordinarily not very highly correlated with the main effects (Althauser, 1971).

This last point is an important one. Consider a 2 × 2 factorial design with equal n. If one uses 0 and 1 coding for the main effect, the resulting coding for interaction is 0, 0, 0, and 1. However, using 1 and −1 codes for the main effects, there is a second set of codes for the interaction, 1, −1, −1, and 1. This second set of codes is uncorrelated with the main effects, whereas the first set has a .289 correlation with each main effect.

EXAMPLE

Sibley (1976) investigated a possible source of bias in the perception of grammaticality of sentences drawn from articles by linguists. She speculated that linguists "set up" the reader to perceive the sentence in the way desired by the linguist. Sibley tested this hypothesis experimentally. The experiment involved 16 linguists and 16 nonlinguists. Half of each received 120 sentences from two articles by prominent linguists in the same order as given by the original author while the other half received the sentences in a random order. The final independent variable was the instructions given to the subjects. Half were told to make an intuitive judgment while the other half were told to use grammatical rules if necessary. The experiment is then a 2 × 2 × 2 factorial design. Four subjects were in each condition and subjects were randomly assigned to order by instruction condition. The dependent variable is the number of "errors" made over the 120 sentences, an error being defined as a disagreement with the author.

In Table 10.3 there is the model matrix for the design. The first dummy, X_1, compares the linguists with the nonlinguists. The second, X_2, compares the author's order with the random order, and X_3 compares the intuition instruction with the rule instruction. Dummy X_4 is formed by multiplying X_1 by X_2; X_5 by X_1 and X_3; X_6 by X_2 and X_3; and X_7 by X_1, X_2, and X_3. These four variables are the interaction effects. All dummies have zero mean and are orthogonal.

In Table 10.4 are the cell means. First note that each dummy variable is a dichotomy and so one can use the formula given in the next chapter to compute the point biserial correlation between each of the dichotomies with the dependent variable Y. For instance, for r_{X_1Y} the formula is

$$\frac{(20.5 - 26.5)\,(.5)}{5.847}$$

yielding −.513. The standard deviation of the dependent variable is 5.847. In Table 10.5 are the other correlations for each of the dummies. Since the dummies are uncorrelated, the correlation is the standardized regression coefficient which in turn is the path coefficient. To find R^2 simply compute the sum of all the correlations squared or .577. In Table 10.5 is the t-test for each coefficient using Equation 4.1. Table 10.5 also contains the unstandardized coefficients, which are identical to the ANOVA effect estimates. Finally, in Figure 10.2 is a path diagram.

The results of Sibley's experiment show that linguists make fewer errors than nonlinguists and that more errors are made when the sentences are in a random order than in the author's order which confirmed Sibley's hypothesis. No interactions are indicated.

DUMMY VARIABLE REGRESSION

There has been recent recognition that the analysis of variance (ANOVA) is only a special case of multiple regression. As has just been

Table 10.3. Model Matrix for Linguistics Study[a]

		$A_1B_1C_1$	$A_1B_1C_2$	$A_1B_2C_1$	$A_1B_2C_2$	$A_2B_1C_1$	$A_2B_1C_2$	$A_2B_2C_1$	$A_2B_2C_2$
X_1:	A	1	1	1	1	−1	−1	−1	−1
X_2:	B	1	1	−1	−1	1	1	−1	−1
X_3:	C	1	−1	1	−1	1	−1	1	−1
X_4:	$A \times B$	1	1	−1	−1	−1	−1	1	1
X_5:	$A \times C$	1	−1	1	−1	−1	1	−1	1
X_6:	$B \times C$	1	−1	−1	1	1	−1	−1	1
X_7:	$A \times B \times C$	1	−1	−1	1	−1	1	1	−1

[a] A_1, linguist; A_2, nonlinguist; B_1, author's order; B_2, random order; C_1, intuition set; C_2, rule set.

Table 10.4. Cell Means for Linguistics Study

	C_1		C_2	
	B_1	B_2	B_1	B_2
A_1	18.00[a]	24.50	16.75	22.75
A_2	24.25	26.25	23.50	32.00

[a]Cell size equals four.

seen by judicious choice of dummy variables, multiple regression can reproduce the results of ANOVA. A relatively straightforward presentation of this process is given in Cohen and Cohen (1975), and a more complicated but highly general presentation is given in Bock (1975).

The virtues of multiple regression over ANOVA are manifold. First, the use of dummy variables encourages researchers to scale their independent variables. Since a scientific law is often best stated as a functional relationship between the independent variable and the dependent variable, the independent variable should be measured on a scale. Often variables in experiments are only operationalized with two levels, high and low, but with multilevel independent variables, the researcher using multiple regression is forced to choose dummy variable weights for the independent variable. Many times there is a natural metric for the independent variable, for example, percent similarity, status, amount of information, or level of motivation. These variables can be easily scaled and the scale weights can be used as a variable and entered into the regression analysis. It can then be tested whether these scaled values explain the variation among the cell means and, more importantly, whether there is no systematic dependent variable varia-

Table 10.5. Analysis of Variance Table[a]

Effect	r and β	Unstandardized Coefficient	t
A	−.513	−3.000	−3.48
B	−.492	−2.875	−3.34
C	−.043	−.250	−.29
A × B	−.043	−.250	−.29
A × C	.171	1.000	1.16
B × C	.128	.750	.87
A × B × C	−.150	−.875	−1.02

[a]R^2 = .577.

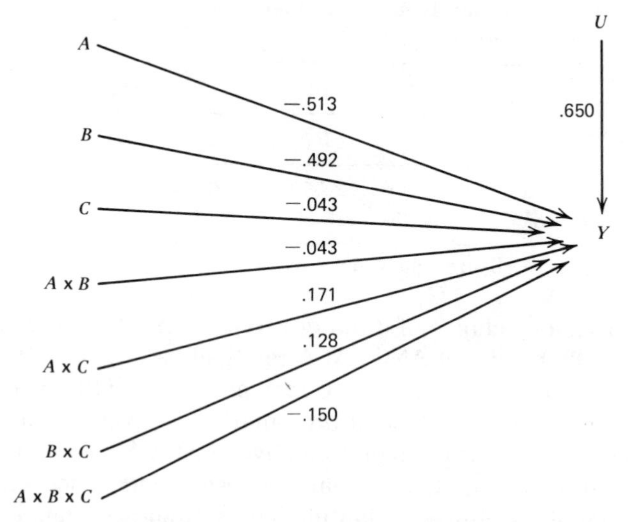

Figure 10.2 Linguistics example.

tion after the scaled independent variable has been controlled. For instance, Byrne and his associates have consistently shown that attraction is a direct function of percent similarity and that there is no evidence of nonlinearity. Byrne can then state not only a relationship between the independent variable and the dependent variable, but he can also state the functional form of the relationship which allows both interpolation and extrapolation of results.

Even if there is no natural scale for the independent variable, one could be derived by the use of scaling methods. One candidate for scaling independent variables might be multidimensional scaling. Levels of an independent variable could be first judged regarding their similarity. Then the similarity judgments could be multidimensionally scaled. The dimensions obtained from the scaling could be used in the dummy coding of independent variables, and then it could be judged which independent variable explained the variation in the cell means.

Another strategy might be to have expert judges scale the independent variable. The cells in a factorial experiment could be judged as to how effective they are in producing change according to a theoretical formulation. "Judged effectiveness" would then be an independent variable in the regression analysis and it would be tested whether it adequately explains the variation among the cell means. Ideally all the systematic variation of the dependent variable should be explained by the scaled independent variable or some simple transformation of the scale (e.g., logarithmic), and the residual effect should be nonsignificant.

Scaling the independent variable provides a dividend of increased power. Since the scaled independent variable is captured by a single variable, it is tested by a single degree of freedom, and Fs with one degree of freedom on the numerator have potentially more power than those with multiple degrees of freedom.

A second advantage of multiple regression over ANOVA is the generality of design. ANOVA is best suited to the case of equal number of subjects in each treatment condition. In fact, ANOVA can be viewed as a simplification in the computations of multiple regression in the equal n case. In the all too common unequal n situation, researchers often employ highly questionable correction procedures. Some researchers randomly discard precious data; others gather data from additional subjects, ignoring the chance that either the subject population or the experimental setting have changed. Equal or unequal sample size presents no special conceptual problems within regression analysis. With unequal n it is true that effects become partially confounded, and the efficiency of effect estimates are reduced. This situation, known as multicollinearity, is not so much a problem but rather the price we pay for unequal n, and of course it is present in either variance or regression analyses.

Given multicollinearity, the between-cells sum of squares cannot be orthogonally partitioned. Unfortunately some researchers have advocated methods that attempt to orthogonally partition the sum of squares: unweighted means analysis and hierarchical model testing. Although the former is an ANOVA strategy and the latter is a multiple regression strategy, both seek equal n ANOVA-like partitioning of variance. The following shows how both unweighted means analysis and hierarchical model testing can both yield anomalous results with unequal n.

Unweighted means analysis assumes that all the cell means are based on the harmonic mean. Unweighted means is meant to be only an *approximation* and is not, as is commonly thought, meant to handle accidental unequal sampling. To illustrate the arbitrariness of the unweighted means analysis, consider the means in Table 10.6. These artificial data indicate that there is a main effect of independent variable B. The values in column 3 have a mean of 13.0 and those in columns 1 and 2 have a mean of 7.0. Confidence in the existence of the main effect should be a function of the distribution of the two sets of sample sizes. The first set of sample sizes in Table 10.6 would inspire greater confidence than the second set of sample sizes because, in the

Table 10.6. Unweighted versus Weighted Means Example

Means

A	B 1	B 2	B 3
1	7	7	13
2	7	7	13

First set of sample sizes

A	B 1	B 2	B 3
1	10	5	10
2	10	5	10

Second set of sample sizes

A	B 1	B 2	B 3
1	10	10	5
2	10	10	5

first set of sample sizes, the deviant means in column 3 are associated with large sample sizes whereas in the second set the deviant means are associated with small sample sizes. Since both have harmonic cell means of 7.5, an unweighted means analysis yields a sum of squares of 360 for the column main effect in both cases. A weighting of means by sample sizes yields a sum of squares for B more in accord with intuition: 432 for the first set of values and 288 for the second set. Unweighted means analysis fails to weight the deviancy of a mean from other means by its sample size.

The second approach to be criticized is called a *hierarchical* approach, or model III by Overall and Spiegel (1969). The effects are first ordered, say A, B, and $A \times B$. First, the effects of A are fitted ignoring the effects of B and $A \times B$. The effect of B is then fitted controlling for A and ignoring $A \times B$. Finally, the $A \times B$ effect is fitted controlling for both A and B. In this way the between-cells sum of squares can be orthogonally partitioned. Note, however, that the effect of A is confounded with B and $A \times B$ and the effect of B with $A \times B$ since the latter effects are not controlled. This fact can create anomalous results. Examine, for instance, the first set of means in Table 10.7. They clearly illustrate only an $A \times B$ effect. The between-group sums of squares is 2500. Using the hierarchical approach, the weighted marginal means for A are 19 and 11. Because of unequal n they are not 15 and 15. Computing sums of squares for A yields 1600. None of the sum of squares can be attributed to B after fitting A. Fitting the interaction controlling for the main effects yields a sum of squares of 900. Thus, the total between-cells sum of squares has been orthogonally partitioned between A and $A \times B$. Counterintuitively, more of the sum of squares is attributed to A than to $A \times B$. This is due to the strong confounding of the A and $A \times B$ effect. Using the dummy codes for A and $A \times B$, the correlation of A with $A \times B$ is .8. Since A explains 64% of the variance of $A \times B$, 64% of the interaction effect is attributed to A (i.e., $.64 \times 2500 = 1600$). Clearly, anomalous conclusions are very possible with the hierarchical mode of model testing. The logic of the hierarchical method implies that if the interaction is significant the main effect should not be tested since the test of the main effect has ignored the significant interaction. In practice, however, most users of the hierarchical approach do not examine the confounding and do not stop testing. I would guess that most users of the hierarchical method do not even know they are using it! The popular MANOVA (Clyde, Cramer, & Sherwin, 1966) program uses a hierarchical method, and the default

Table 10.7. Illustration of Hierarchical Model Testing

First Set			Second Set		
Means					
	B			B	
A	1	2	A	1	2
1	10	20	1	10	18
2	20	10	2	12	20
Sample Sizes					
	B			B	
A	1	2	A	1	2
1	5	45	1	9	15
2	5	45	2	15	9

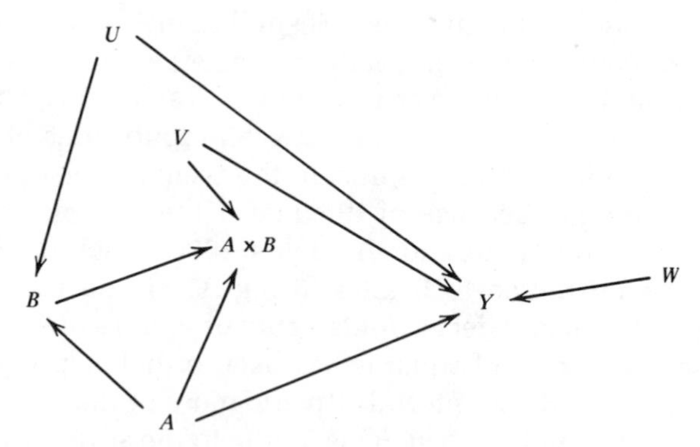

Figure 10.3 Hierarchical ANOVA.

order of testing is simply the order in which the independent variables are read into the program.

It should be noted that ignoring the effects of the other independent variables (i.e., not controlling for them) does not always inflate the sum of squares attributed to an effect. It is possible, although it is empirically unusual, that effects could be grossly underestimated. Given the second set of means in Table 10.7, the marginal means of A are both 15 ignoring B, but clearly there is a main effect of A. This effect is formally identical to the suppressor variable phenomenon in multiple regression.

The hierarchical method implicitly assumes that the independent variables can be causally ordered. For instance, as in Figure 10.3, A causes B and $A \times B$, B causes $A \times B$ and not A, and $A \times B$ does not cause A or B. Then the dependent variable Y is regressed on A, U, and V. The effect for A includes the direct effect of A plus its indirect effect through both B and $A \times B$. The effect of U includes the direct effect of B plus the indirect effect of B through $A \times B$. The effect of V includes only the direct effect of $A \times B$. Thus the direct effect of A could be zero, but A could be judged to have a significant effect if A and B are correlated and B has a direct effect. The procedure yields an orthogonal partitioning of variance since A, U, and V are uncorrelated; however, it does not test effects in a sensible manner. Moreover, the independent variables in experiments have no causal orderings and the notion of indirect effects makes no sense.

Given unequal n, confounding or multicollinearity is inevitable in factorial designs. The solution to the problem is not to find some way to achieve orthogonal partitioning as in the equal n case. Multicollinearity is not a *problem* in experimental design; rather, it is a *cost* that must be paid given the failure to have equal n.

With regression analysis there is a more reasonable approach to the testing of causal effects. The significance of each effect can be judged by comparing the sum of squares with the effect in the model with the sum of squares with the effect removed. All other effects are included in both models. This strategy is called method I by Overall and Spiegel (1969) and has been recommended by a chorus of authors (cf. Overall, Spiegel, & Cohen, 1975). For example, for the first set of means in Table 10.7 no variance would be attributed to the main effects and 900 of the sum of squares would be attributed to interaction. Given confounding, 1600 of the sum of squares cannot be attributed. The reader should also consult an alternative approach given by Applebaum and Cramer (1974).

A third advantage of multiple regression is the ease in pooling interactions into the error term. Anticipated nonsignificant interactions can be dropped from the regression equation, increasing the number of degrees of freedom in the error term and thereby increasing power. This procedure allows for a large number of between-subject independent variables even with a small number of subjects in each cell.

A fourth advantage of the use of multiple regression is that it is flexible as to the type of independent variables. Nominal, ordinal, and interval independent variables are possible in multiple regression. ANOVA is limited to discrete independent variables and analysis of covariance is limited to a set (usually one) of intervally measured independent variables. Some researchers are reluctant to apply multiple regression because of the assumption of linearity. But since multiple regression reproduces the results of ANOVA, the same set of assumptions applies to both. As with ANOVA, multiple regression assumes an additive *model* which is a linear *equation*. Within regression analysis nonlinearity and interaction can be handled by creating additional dummy variables.

ANOVA is in no danger of becoming an endangered statistical species. One major virtue of ANOVA is that it deals with means and the discussion of means often makes greater intuitive sense than regression coefficients. Means can be only indirectly computed from regression coefficients. Multiple regression is often cumbersome to apply: The possible number of terms in the regression equation is the number of cells in the design, and an equation with as many as 64 terms is rarely very comprehensible. The cumbersomeness becomes almost intolerable when the complications of repeated measures and mixed models are added. Many of the features of multiple regression can be applied to ANOVA by the use of special contrasts that are sometimes called planned comparisons. Obviously the researcher should be flexible about when to use ANOVA or multiple regression, but quite clearly multiple regression has some important advantages.

MULTIPLE DEPENDENT MEASURES

Often a researcher is at a loss how to analyze data when there is more than one dependent variable. At the very least, six different strategies are available:

a. Perform repeated measures ANOVA.
b. Perform an ANOVA on the sum of the measures.
c. Perform a MANOVA.
d. Sum the measures weighted by a factor analysis.
e. Covary on one of the measures.
f. Perform a univariate analysis on each of the measures.

The choice among the six strategies can be facilitated by examining three causal models. Imagine two independent variables, X_1 and X_2, and two measured dependent variables, Y_1 and Y_2. In Figure 10.4 path diagrams represent three different causal models for the four variables.

The top causal model in Figure 10.4 assumes that the independent variables cause each of the dependent variables. There are two different approaches to the analysis of this type of model: univariate and multivariate. The univariate strategy would be to perform separate ANOVA's on each of the dependent variables. However, as the number of dependent variables increases, such a strategy would result in a large increase in the probability of both Type I and Type II errors. To reduce the probability of these errors, a multivariate approach should be used. A multivariate F tests the effect of an independent variable on a weighted sum of the dependent variables. The weights for linear combination of the dependent variable are chosen to maximize the univariate F of that linear combination. Significance levels are adjusted since the F is spuriously high. The multivariate F is then "hypothesisless" since the particular nature of the effect is specified by a statistical, not conceptual, criterion.

The middle path diagram in Figure 10.4 contains an unmeasured variable, Y, that the independent variables cause, and Y, in turn, causes

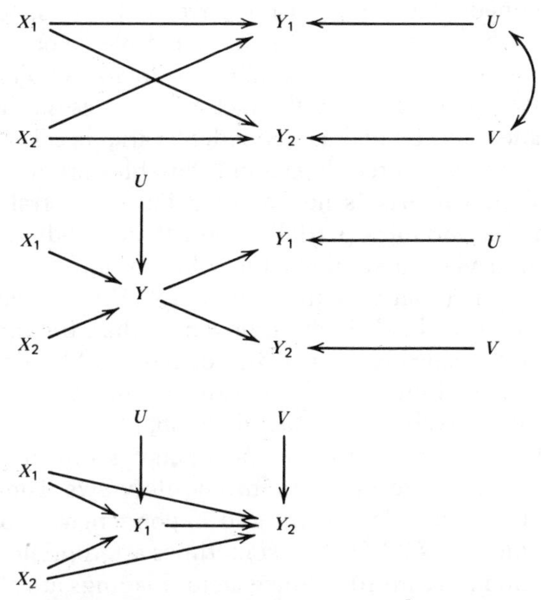

Figure 10.4 Models for repeated measures.

the measured dependent variables. Unlike the top path diagram, the independent variables cause a construct of which the dependent variables are only indicators. An example of this model might be when some manipulation causes arousal which is unmeasured; possible indicators of arousal are fear, tension, and heart rate increase. To test this model, consider two submodels of the middle figure: correlated errors and uncorrelated errors. Correlated errors are represented by the curved line between U and V which is simply a theoretical partial correlation between Y_1 and Y_2 with Y partialled out. If Y totally explains the correlation between Y_1 and Y_2, there are no correlated errors. Given correlated errors, a multivariate analysis is the appropriate method. There should, however, be only one statistically significant solution. As in Figure 9.5 of Chapter 9 only the first root of the canonical analysis should be statistically significant. One should not perform the analysis separately for each independent variable since this would not capture the specification that Y mediate the effects of both independent variables on both dependent variables. One can perform a canonical correlational analysis and then test each independent variable with a step-down F (cf. Bock, 1975); that is, each independent variable is removed from the analysis and the reduction of the canonical correlation is then judged.

If it is assumed that there are no correlated errors, then a factor analytic approach is appropriate. Given no correlated errors, the correlation between the dependent variables is totally explained by the unmeasured variable, that is, by the factor Y. Scores on the factor can then be estimated and treated as dependent variables. To estimate the factor loadings at least three dependent variables are needed. Ideally the raw correlation matrix is not factored but the correlation matrix with independent variables partialled out. If the model is well specified, this should yield a single-factor solution and factor scores can then be estimated. Finally, a univariate ANOVA can be performed on the factor scores. This F is more powerful than the multivariate F since it allows for specification of the construct. More efficient and more general estimation procedures can be obtained by using confirmatory factor analysis, as discussed in Chapter 7, 8, and 9. Alwin and Tessler (1974) used confirmatory factor analysis in an experimental investigation of clients' reactions to initial interviews. They had multiple measures for both independent and dependent variables.

A repeated measures ANOVA is sometimes appropriate for the middle path diagram in Figure 10.4. Since factor loadings are often roughly equal, a simple adding of the variables, as in a repeated measures ANOVA, yields a composite highly correlated with factor scores. Since adding the variables implies an equal weighting, this assumption can be tested by no significant independent variable by repeated measures interaction. To perform a repeated measures ANOVA, all the dependent variables should be measured on the same scale.

The bottom diagram in Figure 10.4 postulates a causal relationship between the dependent variables. Dependent variable Y_1 is assumed to cause Y_2. Variable Y_1 can be a manipulation check, an intervening variable, or a prior measure of Y_2. This model is like the middle diagram but in this case the intervening variable is measured. The appropriate analysis for this model is to compute two univariate ANOVAs. In the first Y_1 is the dependent variable, and in the second Y_1 is a covariate and Y_2 the dependent variable. If one prefers regression analysis, one first regresses Y_1 on X_1 and X_2 and then Y_2 on X_1, X_2, and Y_1. The suggestion of covarying on Y_1 breaks the rule that the covariate should not be caused by the independent variables, but such a rule is important for the main thrust of covariance analysis: increasing power. Covariance analysis can also be used to control for mediating variables like Y_1. One must assume that the reliability of Y_1 is high, say at least .90. (If not, this model becomes the middle diagram in Figure 10.4.) If the reliability of Y_1 is not unity, the effects of X_1 and X_2 on Y_2 are ordinarily though not always overestimated (see Chapter 5). Often for this model it is postulated that Y_1 mediates the causal effect of X_1 and X_2 on Y_2. For instance, Insko et al. (1973) argued that implied evaluation mediates the effects of similarity on attraction. They showed that by covarying implied evaluation out of attraction there was no effect of similarity. This does not demonstrate that the relationship between similarity and attraction is spurious but only that the causal relationship is mediated by an intervening variable, implied evaluation. Serious multicollinearity problems can develop if the relationship between X_1 or X_2 with Y_1 is strong. This would especially be true if Y_1 is a manipulation check. It should be made clear that covarying on a dependent variable is permissible only if it can be assumed that the same variable mediates a causal relationship between the independent variable and another dependent variable.

Thus, a structural modeling approach is useful in determining the mode of statistical analysis. Note that it is not the design *per se* that determines the statistical analysis, but the design plus the causal model. The difficult question is often not the choice of the statistical analysis but the choice of the causal model.

The combination of both between- and within-subjects (repeated measurements) variables can prove to be rather awkward when one sets up a structural model. Nonetheless structural models for such designs can provide richer information than either a repeated measures analysis of variance or multiple regression analysis. For instance, consider a 2 × 2 design with three repeated measurements. Call the within-subjects variable trials and let three dummy variables represent the 2 × 2 between subjects variables: X_1 for the main effect of A, X_2 for the main

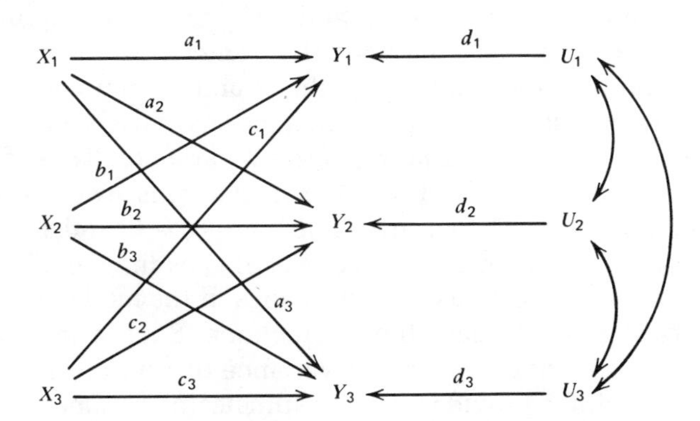

Figure 10.5 A model with between- and within-subject variables.

effect of B, and X_3 for the $A \times B$ interaction. In Figure 10.5 is a path diagram for the six variables. The X and Y variables are *unstandardized,* but the U variables are standardized. There are no curved lines between the X variables because it is assumed that there are an equal number of subjects in each of the four cells of the design; however, if cell sizes were unequal no special problems would arise. The minimum condition of identifiability is met since there are 15 correlations and 15 free parameters, 9 paths, 3 covariances among the X variables, and 3 correlations among the disturbances. Even though the researcher knows that the covariances between X_1, X_2, and X_3 are zero, they still need to be estimated. The model is a hierarchical model and can be estimated by multiple regression and partial correlation analysis.

One can employ LISREL to test the usual analysis of variance hypotheses. For instance, a model in which $c_1 = c_2 = c_3$ tests the $A \times B$ by trials interaction. That is, if there were no interaction the coefficients would be equal. Recall that an interaction specifies that a causal law varies as a function of some other variable. The test of the A by trials interaction sets $a_1 = a_2 = a_3$ and the B by trials interaction sets $b_1 = b_2 = b_3$.

If none of the between-subjects variables interact with trials, then one can proceed to test the A and B main effects and their interaction. They are: the A main effect—$a_1 = a_2 = a_3 = 0$; the B main effect—$b_1 = b_2 = b_3 = 0$; the $A \times B$ interaction—$c_1 = c_2 = c_3 = 0$.

Such an analysis does not make the restrictive assumptions of repeated measures analysis of variance: homogeneity of variance, $d_1 = d_2 = d_3$, and homogeneity of correlation, $\rho_{U_1U_2} = \rho_{U_1U_3} = \rho_{U_2U_3}$. In fact, these assumptions can be tested by LISREL. It is true that the program does assume multivariate normality and the χ^2 test is only approximate; however, it does provide more natural parameter estimates.

As an example of the flexibility of the structural approach, consider the researcher who suspects that the interactions of trials with A, B, and $A \times B$ are an artifact of the scaling of the Y variables. Thus if Y_2 were rescaled by Y_2/k_1 and Y_3 by Y_3/k_2, the interactions would disappear. Such a model presumes that $a_2/a_1 = b_2/b_1 = c_2/c_1 = k_1$ and $a_3/a_1 = b_3/b_1 = c_3/c_1 = k_2$. To test such a model, one estimates the model in Figure 10.6. Note that variable Z has no disturbance. If the model fits the data, then a rescaling can explain the interactions. Such a model can be tested by a multivariate analysis of variance or canonical correlation analysis, but neither provides direct estimates of parameters nor can they allow respecifications such as the Y variables being unmeasured.

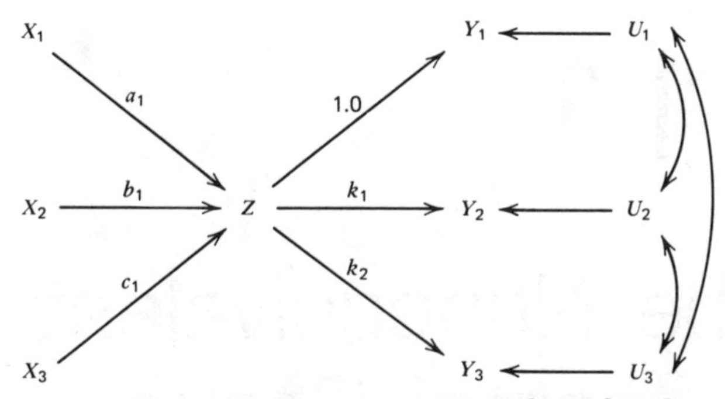

Figure 10.6 Interactions as an artifact of scale.

CONCLUSION

It has been seen that logic of experimental inference can be sharpened by an understanding of the issues of correlational inference. Moreover, even in an experimental context the researcher can employ correlational methods. A structural modeling approach forces researchers to elaborate more complicated causal chains whereas the simple-minded analysis of variance method focuses on a single dependent variable. Social science does not need fewer true experiments but rather experiments that also analyze the data correlationally.

11

The Nonequivalent Control Group Design

A growing concern of social scientists is the evaluation of social programs (Campbell, 1969). Increasingly social scientists are called upon to evaluate educational innovations, behavior modification programs, social work strategies, and job training methods. This chapter shows that structural modeling is equally important in evaluation of social programs as it is in discipline research.

Weiss (1972) has stated that the most common design in evaluation research is a pretest–posttest design in which subjects are not randomly assigned to groups or, as Campbell and Stanley (1963) refer to it, the *nonequivalent control group design*. Thus with this design there is a pretest and posttest, but the control and experimental subjects may differ at the pretest because of the failure to randomize. This initial pretest difference may be because:

- **a.** A treatment is administered to classroom, school, or school system and another classroom, school, or school system is taken as a control group.
- **b.** A true experiment is planned but because of mortality, contamination of the control persons by the experimentals, or variation in the experimental treatment, the true experiment has become a quasi-experiment.
- **c.** Because of scarce resources the treatment is only given to a select group.
- **d.** Subjects select their own treatment level.

Anyone familiar with the problems encountered in natural settings must realize that although randomization and true experimentation are ideal goals in research, they are not always possible. A pretreatment measure does allow the researcher to assess how confounded the experimental treatment is with the subjects' predisposition to respond to the dependent variable.

What follows is a structural analysis of the nonequivalent control group design. The chapter is divided into three sections. The first section specifies a general model. The second section outlines a set of analysis strategies for the design. The third section discusses various special cases of the general model.

GENERAL MODEL

For the nonequivalent control group design there are at least three measured variables: X_1, a pretest measure of the variable that the treatment is to change; X_2, a posttest measure; and T, the treatment variable coded 1 for experimentals and 0 for controls. The causal function of X is divided into three unmeasured, latent components:

G—group difference such as sex, race, classroom, and others,

Z—individual differences within groups,

E—totally unstable causes of X (errors of measurement).

Variable G must be included in the specification of causes of X because in field settings it can rarely be assumed that one samples from a single homogeneous population. Usually multiple populations are sampled with each population having different mean level of X.

For most of what follows all variables, both measured and unmeasured, are standardized because standardization decreases algebraic complexity and because in this case it means little loss of generality. Occasionally, however, the unstandardized measure is employed, as in measures of treatment effect.

The equations for X_1 and X_2 can be expressed in terms of their causes G, Z, and E and any treatment effect:

$$X_1 = a_1G + b_1Z_1 + e_1E_1 \quad [11.1]$$

$$X_2 = a_2G + b_2Z_2 + tT + e_2E_2 \quad [11.2]$$

where the subscripts 1 and 2 refer to time. It is assumed that the group differences variable, G, is perfectly stable making its autocorrelation unity which explains why G needs no time subscript. Relative position within groups, Z, may not be perfectly stable making its autocorrelation less than one, while errors of measurement, E, are perfectly unstable making its autocorrelation zero. (Uncorrelated measurement errors are assumed by only one analysis strategy discussed here: analysis of covariance with reliability correction.) It is also assumed that all unmeasured variables are uncorrelated with each other with the previously stated exception that Z_1 and Z_2 may be correlated ($\rho_{Z_1Z_2} = j$).

If the treatment is correlated with the pretest, it must then be confounded with the causes of the pretest. Thus, the treatment must be correlated with group differences, relative position within groups, errors of measurement, or any combination of the three. So in writing the structural equation for the treatment variable, the variables G, Z, and E must be included. It is assumed that the occasion of selection of the persons into treatment occurs at the pretest, thus making T confounded with Z and E at time one. The causal function of the treatment variable is then

$$T = qG + mZ_1 + sE_1 + fU \qquad [11.3]$$

where U is a residual term that is uncorrelated with all the other variables in the model and is simply a variable that represents all other causes of selection besides G, Z, and E.

Equations 11.1, 11.2, and 11.3 have been expressed in the path diagram in Figure 11.1. The correlations between the measured variables are:

$$\rho_{X_1T} = qa_1 + mb_1 + se_1 \qquad [11.4]$$

$$\rho_{X_2T} = qa_2 + mjb_2 + t \qquad [11.5]$$

$$\rho_{X_1X_2} = a_1a_2 + b_1b_2j + t\rho_{X_1T} \qquad [11.6]$$

The treatment variable has been dummy coded, 1 for treatment and 0 for control. A correlation of a dichotomous variable with another variable is sometimes called a point biserial correlation and is itself a product moment correlation. Campbell (1971) has suggested that these *treatment–effect correlations* are very useful in assessing treatment effects. A simple correlation may seem to be an unusual measure of difference between treated groups, but it can be shown that both mean difference and the t ratio can be easily transformed into r. The formula for t to r is

$$r = \frac{t}{(t^2 + N-2)^{1/2}}$$

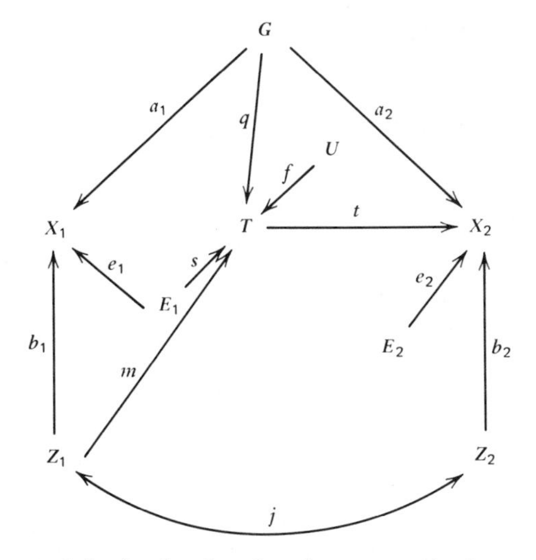

Figure 11.1 Model of selection for the nonequivalent control group design.

where N is the total number of subjects. The formula for the difference between treated mean ($\overline{X}_T$) and control mean ($\overline{X}_C$) to r is

$$r = \frac{(\overline{X}_T - \overline{X}_C)\ (P_T P_C)^{1/2}}{s_X}$$

where s_X is the variability of the observations, P_T the proportion of the total sample in the treated group, and P_C the control group ($P_T + P_C = 1$). The treatment–effect correlation is then closely related to t and the difference between means, but like all measures of correlation it is a standardized measure of the strength of a relationship. The use of dummy codes for categorical independent variables is discussed in the previous chapter.

ANALYSIS STRATEGIES

Designate μ_{E1} as the population mean of the experimental group at the pretest, and μ_{C1} as the control group mean at the pretest. Both μ_{E2} and μ_{C2} refer to means at the posttest. Those means now refer to the *unstandardized* pretest and posttest. Ordinarily, in the nonequivalent control group design there is an initial difference such that $\mu_{E1} - \mu_{C1}$ is nonzero. These pretreatment differences then make the interpretation of any difference between groups on the posttest problematic. At issue is whether pretreatment differences should increase, decrease, or remain stable given no treatment effect.

The literature suggests four modes of statistical analysis to measure treatment effects in the nonequivalent control group design:

a. Analysis of covariance.
b. Analysis of covariance with reliability correction.
c. Raw change score analysis.
d. Standardized change score analysis.

In Table 11.1 the differences between these four techniques can be illustrated by considering what each technique takes as the dependent variable. In covariance analysis the pretest is a covariate and the residual from the posttest–pretest regression is the dependent variable. The results of analysis of covariance are virtually equivalent to a multiple regression analysis in which the pretest and treatment variable are the predictors, and the posttest the criterion. Both Lord (1960) and Porter and Chibucos (1974) have suggested that the estimated regression coefficient is attenuated by measurement error in the pretest, and so the regression coefficient must be corrected for attenuation. For raw change score analysis the pretest is simply subtracted from the posttest. Although change score analysis is often condemned (e.g., Cronbach & Furby, 1970; Werts & Linn, 1970), it is perhaps the most common way of analyzing this design. The popularity of this mode of analysis is no doubt due to its seeming ease of interpretation and the fact that it can be viewed as part of a repeated measures analysis of variance. The main effect for treatment in raw change score analysis is equivalent to the Time by Treatment interaction of a repeated-measures analysis of variance (Huck & McLean, 1975). It should be noted that using the change score as the dependent variable and the pretest as a covariate yields significance test results that are identical to the analysis of covariance with a metric of change (Werts & Linn, 1970). Another mode of analysis is standardized change score analysis. As is seen later, this is actually a correlational method, but it is possible to view the dependent variable of the analysis as the posttest minus the pretest multiplied by the ratio of time 2 pooled within treatment groups standard deviation to the time 1 pooled within treatment groups standard deviation. Kenny (1975*a*) presents the dependent variable as the difference between the standardized (unit variance, zero mean) pretest and posttest. For reasons to be discussed later, the dependent variable in Table 11.1 is preferable.

Table 11.1. Dependent Variables and Null Hypotheses for Four Modes of Analysis Where X_1 is the Pretest, X_2 the Posttest, and T the Treatment Variable[a]

Statistical Technique	Dependent Variable	Measure of Treatment Effect	Correlational Null Hypothesis
Analysis of covariance	$X_2 - b_{21 \cdot T} X_1$	$(\bar{X}_{2T} - \bar{X}_{2C}) - b_{21 \cdot T}(\bar{X}_{1T} - \bar{X}_{1C})$	$\rho_{X_2T} - \rho_{X_1T}\rho_{X_1X_2} = 0$
Covariance with reliability correction	$X_2 - b^*_{21 \cdot T} X_1$	$(\bar{X}_{2T} - \bar{X}_{2C}) - b^*_{21 \cdot T}(\bar{X}_{1T} - \bar{X}_{1C})$	$\rho_{X_2T} - \rho_{X_1T}\rho_{X_1X_2}/\rho_{X_1X_1} = 0$
Raw change score analysis	$X_2 - X_1$	$(\bar{X}_{2T} - \bar{X}_{2C}) - (\bar{X}_{1T} - \bar{X}_{1C})$	$\rho_{X_2T} - \rho_{X_1T}\sigma_{X_1}/\sigma_{X_2} = 0$
Standardized change score analysis	$X_2 - (s_{X_2 \cdot T}/s_{X_1 \cdot T})X_1$	$(\bar{X}_{2T} - \bar{X}_{2C}) - s_{X_2 \cdot T}/s_{X_1 \cdot T}(\bar{X}_{1T} - \bar{X}_{1C})$	$\rho_{X_2T} - \rho_{X_1T} = 0$

[a] $b_{21 \cdot T}$ is the unstandardized partial regression of X_2 on X_1, controlling for T, and * indicates corrected for attenuation.

There is for each mode of analysis a statistical expression that equals zero if there were no treatment effects, that is, a null hypothesis for each method. Discussed are two different ways of expressing this null hypothesis: means and treatment–effect correlations. Although both ways are algebraically equivalent, each has its own distinct advantage. The use of means gives a measure of treatment effect in the metric of the posttest. Treatment–effect correlations are useful in determining the relative difference between the experimental and control group, that is, relative to the standard deviation.

For each of the four modes of analysis the measure of treatment effect can be viewed as the difference between means at the posttest ($\mu_{E2} - \mu_{C2}$) minus the difference at the pretest ($\mu_{E1} - \mu_{C1}$) times some correction factor. Porter and Chibucos (1974) call this correction factor the *index of response*. In Table 11.1 are the measures of treatment effect for each of the four methods. For analysis of covariance the correction factor is $b_{21 \cdot T}$, the unstandardized partial regression coefficient of X_2 on X_1 controlling for T. For analysis of covariance with reliability correction the correction factor is $b_{21 \cdot T}$ corrected for attenuation due to measurement error. The appropriate reliability is the pooled within-treatment groups reliability. Raw change score analysis has no correction factor or, more precisely, a correction factor of one. For standardized change score analysis the correction factor is $s_{X_2 \cdot T}/s_{X_1 \cdot T}$ where $s_{X_2 \cdot T}$ is the pooled within-treatment standard deviation of the posttest.

All four null hypotheses can also be viewed as setting the posttest–treatment correlation equal to the pretest–treatment correlation times a correction factor. For analysis of covariance the correction factor is the pretest–posttest correlation. Since the pretest–posttest correlation is ordinarily less than one, analysis of covariance implies that in the absence of treatment effects, the posttest–treatment correlation is less in absolute value than the pretest–treatment correlation. The correction factor of the analysis of covariance with reliability correction is the pretest–posttest correlation divided by the reliability of the pretest. This correction factor is ordinarily less than one, making posttest–treatment ordinarily less in absolute value than the pretest–treatment correlation. The correction factor for raw change score analysis is the standard deviation of the pretest divided by the standard deviation of the posttest. The correction factor for standardized change score analysis is equal to one.

To demonstrate the treatment–effect correlation measures of treatment effects in Table 11.1, take the covariance between the treatment with the dependent variable and simplify the expression. For analysis of covariance the covariance of the treatment with the dependent variable is

$$C(T,X_2 - b_{X_2X_1 \cdot T}X_1)$$
$$C(T,X_2) - b_{X_2X_1 \cdot T}C(T,X_1)$$

Since

$$b_{X_2X_1 \cdot T} = \frac{C(X_1,X_2) - C(X_1,T)C(X_2,T)/\sigma_T^2}{\sigma_{X_1}^2 - C(T,X_1)^2/\sigma_T^2}$$

then

$$C(T,X_2) - \frac{C(T,X_1)C(X_1,X_2) - C(X_1,T)^2C(X_2,T)\sigma_T^2}{\sigma_{X_1}^2 - C(T,X_1)^2/\sigma_T^2}$$

Multiplying through by

$$\sigma_{X_1}^2 - \frac{C(T,X_1)^2}{\sigma_T^2}$$

and subtracting yields

$$\sigma_{X_1}^2C(T,X_2) - C(T,X_1)C(X_1,X_2)$$

Dividing through by $\sigma_T\sigma_{X_2}\sigma_{X_1}^2$ yields

$$\frac{C(T,X_2)}{\sigma_T\sigma_{X_2}} - \frac{C(T,X_1)}{(\sigma_T\sigma_{X_1})}\frac{C(X_1X_2)}{(\sigma_{X_1}\sigma_{X_2})}$$

$$\rho_{TX_2} - \rho_{TX_1}\rho_{X_1X_2}$$

The preceding formula is the numerator of the formula for $\rho_{X_2T \cdot X_1}$ and $\beta_{X_2T \cdot X_1}$, that is, the partial correlation and standardized partial regression coefficient of X_2 and T controlling for X_1.

The logic for analysis of covariance with reliability correction is the same as previously with the inclusion of the reliability of the pretest in the formula.

For raw change score analysis the covariance is

$$C(T,X_2 - X_1)$$
$$C(T,X_2) - C(T,X_1)$$

Multiplying through by $\sigma_{X_1}/(\sigma_{X_1}\sigma_{X_2}\sigma_T)$, yields

$$\frac{C(T,X_2)}{\sigma_T\sigma_{X_2}} - \frac{(\sigma_{X_1}/\sigma_{X_2})C(T,X_1)}{\sigma_T\sigma_{X_1}}$$

$$\rho_{TX_2} - \left(\frac{\sigma_{X_1}}{\sigma_{X_2}}\right)\rho_{TX_1}$$

For standardized change score analysis

$$C\left(T, X_2 - \left(\frac{\sigma_{X_2 \cdot T}}{\sigma_{X_1 \cdot T}}\right)X_1\right)$$

$$C(T,X_2) - C(T,X_1)\frac{\sigma_{X_2}(1-\rho_{X_2T}^2)^{1/2}}{\sigma_{X_1}(1-\rho_{X_1T}^2)^{1/2}}$$

Multiplying through by $1/(\sigma_T\sigma_{X_2})$ yields

$$\rho_{TX_2} - \rho_{TX_1}\frac{(1-\rho_{X_2T}^2)^{1/2}}{(1-\rho_{X_1T}^2)^{1/2}}$$

The preceding equals zero except in trivial cases only if $\rho_{TX_2} - \rho_{TX_1}$ equals zero.

Campbell and Erlebacher (1970) have suggested a fifth method: covariance analysis with common factor coefficient correction. This method closely resembles analysis of covariance with reliability correction, the difference being that the regression coefficient is divided by the pretest–posttest correlation of X instead of the reliability of X_1. This correction yields the same correlational null hypothesis as that for standardized change score analysis since the pretest–posttest correlations cancel each other.

It should be clear from Table 11.1 that the measures of treatment effect of the four modes of analysis are different hypotheses except in highly trivial cases. Each of the four modes of analysis has been advocated as *the* method of analysis for the nonequivalent control group design by various authors. Other authors (e.g., Lord, 1967) have pointed out that it is paradoxical that different methods yield different conclusions. Cronbach and Furby (1970) state that treatments cannot be compared for this design. The literature on this design is, therefore, very confusing and not at all instructive to the practitioner.

SPECIAL CASES OF THE GENERAL MODEL

The validity of any mode of analysis depends on its match with the process of selection into groups. In the remaining part of this chapter, various special cases of the general model of selection are considered. Each mode of analysis is appropriate for a given model of selection. These special cases of the general model yield overidentifying restrictions given no treatment effects. It will then be shown that these overidentifying restrictions derived from the model of selection match the null hypothesis of each of the four modes of analysis given in Table 11.1.

Three different types of selection processes are considered. The first type is selection based on the measured pretest. This is the only type of selection considered for which selection is controlled by the researcher. In this case either by design or accident the treatment is correlated with the entire pretest and, therefore, correlated with all the causes of the pretest: G, Z, and E. For this type of selection process the analysis of covariance is appropriate. The second type of selection is selection based on the true pretest. For this case the treatment is related to G and Z and not E, making the analysis of covariance with reliability correction appropriate. The third type of selection is based on group differences. Subjects are assigned to the treatment because of demographic and social variables, G. Given that the effect of these demographic variables is stationary over time, standardized change score analysis is the appropriate mode of analysis. Finally discussed is the case in which the occasion of selection into treatment groups is not at the pretest.

Selection Based on the Pretest

At times it may be possible to control selection into treatment groups, but it is decided not to randomly assign subjects to treatment conditions. One reason for not randomly assigning is that it may not be fair for all types of subjects to have the same probability of receiving the treatment, that is, certain persons (e.g., the injured or disadvantaged) are considered more deserving of the treatment. One strategy, called the regression discontinuity design by Thistlewaite and Campbell (1960), is to assign subjects to the treatment on the basis of pretest. Persons scoring above (below) a certain point would be given the treatment and those scoring below (above) or equal to that point would be the controls. (Since the pretest is measured on an interval scale and treatment is dichotomous, ρ_{X_1T} does not equal one but it is high.) Actually a pretest itself need not be used, but rather any measure of "deservingness." The treatment, then, is a function of the pretest (X_1) and random error (U):

$$T = kX_1 + fU \qquad [11.7]$$

In this case assignment to the treatment groups is deliberately confounded with the pretest and, therefore, with the causes of the pretest while in the true experimental case, the treatment is deliberately uncorrelated with the pretest through randomization. But like the true experimental case, a specification has been gained by controlling the assignment to groups. In the population the treatment should correlate with the unmeasured causes of pretest (G, Z_1, and E_1) to the degree to which they cause the pretest. Substituting Equation 11.1 for X_1 with Equation 11.7 yields

$$T = k(a_1G + b_1Z_1 + e_1E_1) + fU$$

Taking the covariance of G, Z_1, and E_1 with the preceding equation yields a_1k, b_1k, and e_1k. Since one can also solve for the same covariances from Equation 11.3, it follows that

$$a_1k = q \qquad [11.8]$$

$$b_1k = m$$

$$e_1k = s$$

or alternatively

$$\frac{q}{a_1} = \frac{m}{b_1} = \frac{s}{e_1} = k$$

Given Equations 11.4 and 11.8 it follows that

$$\rho_{X_1T} = a_1(a_1k) + b_1(b_1k) + e_1(e_1k)$$

$$= k(a_1^2 + b_1^2 + e_1^2)$$

Since $a_1^2 + b_1^2 + e_1^2 = 1$ through standardization, it follows that

$$\rho_{X_1T} = k \qquad [11.9]$$

If no treatment effects are assumed ($t = 0$), given Equations 11.5 and 11.8 it follows that

$$\rho_{X_2T} = a_2(a_1k) + b_2j(b_1k)$$

$$= k(a_1a_2 + b_1b_2j)$$

Given no treatment effects and Equation 11.6, it follows that

$$\rho_{X_2T} = k\rho_{X_1X_2} \qquad [11.10]$$

Solving both Equations 11.9 and 11.10 for k, it then follows that

$$\rho_{X_2T} - \rho_{X_1T}\rho_{X_1X_2} = 0$$

which is the null hypothesis for analysis of covariance in Table 11.1. Thus, when subjects are assigned to the treatment on the basis of the pretest, analysis of covariance is the appropriate mode of analysis.

Rubin (1977) and Overall and Woodward (1977) have shown that much more complicated decision rules than regression discontinuity can be used to determine selection and covariance analysis can still be valid. However, the researcher must still control selection into treatment groups. Special attention should be paid to the validity of covariance's assumptions of homogeneity and linearity of regression. Lack of linearity can usually be remedied by including higher-order terms such as the pretest squared. Given the little or no overlap in the pretest distributions of the treated and untreated groups, violations of linearity could lead to erroneous conclusions and may be indistinguishable from a treatment by pretest interaction.

It seems to be an all too common occurrence that randomization of subjects into treatment groups produces treatment differences even before the treatment is administered. Although one realizes that the probability of such pretest differences is 1 out of 20, given the conventional .05 level of significance, there is nonetheless the feeling of being persecuted by fate. It seems that the experiment has been doomed and there is no way to achieve valid inference.

Unhappy randomization does not mean cases of failure to randomize or cases in which there is randomization but a selected group of control and experimental subjects fail to provide posttest data. It is a randomized experiment with a pretest difference between the experimentals and controls. Randomization has not failed, as it is sometimes mistakenly thought; only an unlikely type of event has occurred.

Valid inference is possible in a way similar to our discussion of the regression discontinuity design. If there is a pretest difference, the treatment is confounded with the causes of the pretest. The *expected* degree of this confounding with each cause is proportional to its causal effect on the pretest. For a *randomized experiment* with the pretest correlated with the treatment, the analysis of covariance is not only appropriate but necessary.

Selection Based on the Pretest True Score

The main concern here is not for when the researcher controls selection into treatment groups but for when this is not the case. Two types of subject selection are discussed: first, selection based on the pretest true score (G and Z) and, second, selection based only on group differences (G).

The model previously discussed assumes that the treatment is correlated with the errors of measurement in the pretest. This is generally implausible when selection is uncontrolled. If persons select themselves into programs, the variables that determine selection into the treatment are likely to be correlated with true ability and not chance performance on the pretest. The subjects, not the experimenter, control selection making the treatment correlated with only the true causes of the pretest. These true scores are not actually measured but the causes of selection like motivation, expectation, and encouragement are correlated with true ability. As earlier, it is assumed that the ratio of the effect of group differences on the treatment to its effect on the pretest equals the ratio of the effect of individual differences on the treatment to its effect on the pretest; that is,

$$\frac{q}{a_1} = \frac{m}{b_1} = k \qquad [11.11]$$

This assumption presumes that selection into treatment groups is on the basis of the true pretest, $a_1G + b_1Z_1$, and not on errors of measurement or any other function of G and Z. In the case of selection based on the measured pretest, a similar hypothesis was made but that hypothesis was justified by the design of the research. In this case Equation 11.11 is an assumption that must be justified by evidence from the selection process itself.

If it is assumed that $s = 0$ and Equation 11.11 holds, correlations 11.4 and 11.5 become

$$\rho_{X_1T} = k(a_1^2 + b_1^2) \qquad [11.12]$$

$$\rho_{X_2T} = k(a_1a_2 + b_1b_2j) + t$$

Since the reliability of the pretest $\rho_{X_1X_1}$ is defined as $a_1^2 + b_1^2$, it follows $\rho_{X_1T} = k\rho_{X_1X_1}$. Also given no treatment effects and Equation 11.6, it follows that $\rho_{X_2T} = k\rho_{X_1X_2}$. The analysis of covariance null hypothesis will not equal zero since

$$\rho_{X_2T} - \rho_{X_1T}\rho_{X_1X_2} = k\rho_{X_1X_2} - (k\rho_{X_1X_1})\rho_{X_1X_2}$$

$$= k\rho_{X_1X_2}\,(1-\rho_{X_1X_1})$$

Except in trivial cases, the preceding will only equal zero if the reliability of the pretest is perfect.

The reason for this bias is that the within groups regression coefficient is attenuated because the pretest is measured with error. The posttest should be regressed not on the *measured* pretest as in covariance analysis, but on the *true* pretest (see Chapter 5). It is because of this bias that both Lord (1960) and Porter and Chibucos (1974) have suggested a correction for the analysis of covariance. One can view this correction as correcting the regression coefficient for unreliability in the pretest. To make the reliability correction we must have an estimate of the reliability of the pretest. Assuming that there is an estimate of reliability and $s = 0$ and Equation 11.11 holds, the analysis of covariance with reliability correction yield unbiased estimates of treatment effects.

Porter and Chibucos (1974) discuss in detail computational methods for true score correction. Regress the pretest on the treatment variable to obtain b_{X_1T}. Now compute

$$X^A = b_{X_1T}T + \rho_1(X_1 - b_{X_1T}T)$$

where ρ_1 is the within treatment group reliability of the pretest and X^A is the adjusted pretest. Now regress X_2 on X^A and T, and the regression coefficient for T is unbiased if s equals zero and Equation 11.11 holds.

A difficulty with the reliability correction procedure is the necessity of having a reliability estimate. The inclusion of any ad hoc estimate, for example, internal consistency, into the correction formula almost certainly increases the standard error of the estimate. Ideally the reliability estimation procedure should be part of a general model. Following Lord (1960) consider, for example, a parallel measure of X_1, say Y. Let

$$Y = a_3G + b_3Z_1 + e_3E_3$$

where E_3 is uncorrelated with all other unmeasured variables. If it is assumed Equation 11.11 holds and that $t = s = 0$, then it follows that

$$\rho_{X_1Y} = a_1a_3 + b_1b_3 \qquad [11.13]$$

$$\rho_{X_2Y} = a_2a_3 + b_2b_3j$$

$$\rho_{TY} = k(a_1a_3 + b_1b_3) \qquad [11.14]$$

Given Equations 11.12, 11.13, and 11.14, it follows that

$$\rho_{X_1X_1} = \frac{\rho_{TX_1}\rho_{X_1Y}}{\rho_{TY}}$$

Substituting the preceding formula for reliability into the reliability correction formula in Table 11.1 yields the following null hypothesis:

$$\rho_{X_2T} = \frac{\rho_{TY}\rho_{X_1X_2}}{\rho_{X_1Y}}$$

or equivalently in vanishing tetrad form:

$$\rho_{X_2T}\rho_{X_1Y} - \rho_{TY}\rho_{X_1X_2} = 0$$

The vanishing tetrad can be tested by the null hypothesis of a zero second canonical correlation between variables X_1 and T and variables X_2 and Y as in Chapter 7. Note that it has been assumed that the true score of X_1 selected persons into treatment, and not the true score of Y. However, if both true scores were identical, it follows that

$$\frac{a_1}{a_3} = \frac{b_1}{b_3}$$

and that

$$\rho_{X_1X_2}\rho_{YT} - \rho_{X_1T}\rho_{YX_2} = 0$$

Selection Not at the Pretest

It has been assumed that the occasion for selection into treatment is at the pretest. For most programs the occasion for selection is not so well defined. Treated and control subjects may drop out of the program or move out of the area, some controls may enter the program, unsuccessful or unhappy treated subjects may not show up at the posttest, and so on. Sometimes the "pretest" takes place well before the treatment begins; for example, test scores of the previous year are used as a pretest for a remedial program for the current year. For many real world programs the occasion of selection into the program is not identical with the pretest.

For assignment based on the true score, the correlation of the treatment with the effect variable is not the highest at the pretest. This would substantially bias analysis of covariance with reliability correction if selection does not occur at the pretest. Imagine that the researcher does not measure X_1 but uses X_0 as a "pretest." For such a case the analysis of covariance with reliability correction would ordinarily be biased. For instance, if parameters a, b, and j remained stationary over time, then the treatment–effect correlations are equal to each other, given no treatment effects. Thus, if selection occurs midway between the pretest and posttest, or "averages out" midway, and if the researcher can assume stationarity, then standardized change score analysis is the appropriate form of analysis. If selection is based on group differences, the occasion of selection is irrelevant since group differences are perfectly stable.

Selection on the Basis of Group Differences

For most social programs, assignment to the treatment is not based on some psychological individual difference, that is, true score, but on some sociological, demographic, or social psychological characteristic. This "sociological" selection is brought about in a variety of ways:

a. It may be a matter of policy or legislation that treatment is available to a particular social group, for example, persons living in particular census tracts. It may be virtually impossible to find some members of that social group who did not receive the treatment.

b. Some treatments are administered to members of an entire organization, for example, school system, and members of the treated organization must be compared with another organization.

c. To receive a selective treatment a person or their sponsor must be highly motivated, or have political connections and organizational "savvy." These volunteers differ systematically from nonvolunteers on a number of characteristics (Rosenthal & Rosnow, 1975).

d. Sometimes the treatment either is a sociological or demographic variable or hopelessly confounded with one. Examples of this are a study on the effects of dropping out of high school and testing for differences in socialization between males and females.

The reader can probably also conceive other patterns of sociological selection. Suffice it to say, it is a rather common form of selection into social programs.

In terms of the general model, selection based on group differences implies that $s = m = 0$. However, treatment effects are still not identified if they exist, and consequently there is no overidentifying restriction given the absence of treatment effects. To be able to estimate treatment effects one must take additional assumptions or add on additional measures. Following Kenny (1975*a*), it is shown in this chapter that the fan spread hypothesis can allow for identification.

One way to gain an overidentifying restriction is to assume some form of stationarity, that is, assume that the effect of group differences is the same at the pre- and posttest. Campbell (1967) has argued for just such a model with what he called the "fan spread hypothesis." The hypothesis is that associated with the mean differences between groups is a difference in maturation: Those with the higher mean mature at a greater rate than those with the lower mean. Campbell calls this the "interaction of selection and maturation" and has used this interaction as an argument against raw change score analysis. Since the mean difference between groups is widening over time, change score analysis indicates only the more rapid rate of maturation of the initially higher group. The fan spread hypothesis is that accompanying increasing mean differences is increasing variability within groups. In its strictest form the fan spread hypothesis is that the difference between group means relative to pooled standard deviation within groups is constant over time.

The rationale for the fan spread hypothesis is that the different groups are members of different populations living in different environments. The different environments create and maintain different levels of performance and different rates of growth. Given that growth is a cumulative process, variability increases over time. The groups would eventually asymptote at different levels.

Formally, the model for selection on the basis of group differences is

$$T = qG + fU$$

Fan spread and no treatment effects implies that $a_1 = a_2$, making $\rho_{X_1T} = \rho_{X_2T}$, that is, equality of the treatment–effect correlations.

Recall that fan spread is only a hypothesis. It may not hold in any particular data set. Ideally fan spread should be tested internally by the data itself (Kenny & Cohen, 1978). Alternative specifications other than fan spread are also possible (Bryk & Weisberg, 1977; Linn & Werts, 1977).

If the variance within treatment groups of the pretest and posttest is stationary over time, then the results of standardized change score analysis and raw change score analysis converge. Thus, for cases in which variance is stationary, raw change score analysis is to be preferred since standardization is needed only to stabilize variance over time. Power transformations may also be applied to stabilize within treatment variance, in a way akin to meeting the homogeneity of variance assumptions of analysis of variance.

It is not advisable to actually standardize the pretest and posttest and perform a standardized change score analysis as was implied in Kenny (1975*a*), since the standard deviation is itself affected by any treatment effect. Rather, one should adjust the pretest by $s_{X_2 \cdot T}/s_{X_1 \cdot T}$ and then perform a change score analysis. As defined earlier, $s_{X_2 \cdot T}$ is the standard deviation of the posttest with the variance of the treatment partialled out, that is, in the case of discrete treatment groups, the pooled within group treatment standard deviations. A difficulty with using this adjustment is that sampling variation is surely introduced since $s_{X_2 \cdot T}/s_{X_1 \cdot T}$ is a sample statistic, not a population value. Nonetheless, using $X_2 - (s_{X_2 \cdot T}/s_{X_1 \cdot T})X_1$ as the dependent variable yields an interpretable, simple metric in the posttest.

One might also simply compare treatment–effect correlations over time. This has the disadvantage that an interpretable measure of treatment effects must then be derived. It does, however, have the advantage that the summary inferential statistic does not vary according to which score is taken as the pretest and which is the posttest. Note that the statistical test results differ if $X_2 - (s_{X_2 \cdot T}/s_{X_1 \cdot T})X_1$ is the dependent variable or if $X_1 - (s_{X_1 \cdot T}/s_{X_2 \cdot T})X_2$ is the dependent variable.

The test of significance between two treatment–effect correlations is the Hotelling-Williams test (Williams, 1959). It is

$$t_{(N-3)} = \frac{(r_{X_2T} - r_{X_1T})((N-1)(1 + r_{X_1X_2}))^{1/2}}{(2d(N-1)/(N-3) + (r_{X_1T} + r_{X_2T})^2(1 - r_{X_1X_2})^3/4)^{1/2}}$$

where d equals

$$1 - r_{X_1T}^2 - r_{X_2T}^2 - r_{X_1X_2}^2 + 2r_{X_1T}r_{X_2T}r_{X_1X_2}$$

and N is the sample size.

It is more likely that mean differences are proportional to the true standard deviations as opposed to the measured standard deviations. Thus the index of response should be $(\sigma_{X_2}/\sigma_{X_1})\ (\rho_{X_2X_2}/\rho_{X_1X_1})^{1/2}$ where $\rho_{X_2X_2}$ is the reliability of the posttest and $\rho_{X_1X_1}$ the reliability of the pretest. All standard deviations and reliabilities are defined within treatments. Note that if $\sigma_{X_2}/\sigma_{X_1}$ is used as the index of response it must then be assumed that $\rho_{X_1X_1} = \rho_{X_2X_2}$.

Before moving on to consider an example it should be pointed out that all the statistical techniques can be applied in a multivariate fashion. That is, if there are background variables like age, sex, and ethnicity, they can be entered into the analysis since all four of analysis techniques can be viewed as application of multiple regression. For analysis of covariance the posttest is the criterion and the pretest, treatment, and background variables are the predictors. For the analysis of covariance with reliability correction the pretest must be adjusted before it is entered into the prediction equation. The pretest is regressed on the treatment and background variables, $\hat{X}_1$. Then $\rho(X_1 - \hat{X}_1)$

$+ \hat{X}_1$ is entered into the regression equation (where ρ is the reliability of $X_1 - \hat{X}_1$) along with the treatment and background variables. For raw change score analysis the criterion is simply raw change and the treatment and background variables are the predictor variables. For standardized change the criterion is the posttest minus the pretest times the standard deviation of the posttest with the treatment *and background variables* partialled out divided by the pretest standard deviation again with the treatment and background variables partialled out. The predictor variables are again the treatment and background variables.

The treatment need not be conceived of as a simple dichotomy. It may be a multilevel variable, and multiple dummy variables are then formed. Moreover, the treatment variable may be intervally scaled as in the number of treatment hours received and treatment–background variable interactions may be entered.

To compute treatment–effect correlations, one needs to compute partial treatment–effect correlations. That is the background variables, and other treatment variables should be partialled out.

Example

Steven M. Director (1974) reanalyzed an evaluation of a manpower training program. The interesting aspect of the design is that there are two *pre*treatment measures of yearly income, the measures being separated by one year. The analyses in this example use the two pretest measures as the pretest and posttest. Since the treatment had not been administered, the analysis should indicate no effect.

In Table 11.2 are the basic summary statistics. Note that those who received the manpower training started lower than the control group

Table 11.2. Means and Standard Deviations in Dollars of the Manpower Training Evaluation

	Year	
Group	**1966**	**1967**
Experimental	2024	2744
(N = 181)	(2294)[a]	(2719)
Control	3140	3994
(N = 1827)	(2574)	(2767)

[a]Standard deviation.

and remained lower. In fact the gap actually widened by $134.00. The pretest–posttest correlation between income is .808.

In Table 11.3 are the indices of response for four methods of analysis. There is no reliability measure for the pretest, but for illustrative purposes it has been set to a .90 value. Note that all four estimates of the nonexistent treatment are negative, indicating that the phantom program reduced income. Since the program is compensatory, the larger the index of response, the less negative the estimate of treatment effect. Standardized change gives the least negative estimate of effect and covariance the largest. Only the covariance estimate reaches statistical significance. Although this example makes covariance look bad, I could have just as easily chosen an example that makes standardized change analysis look bad.

A critical issue has been the direction of bias in analysis of covariance when it is misapplied. Campbell and Erlebacher (1970) have stated that covariance tends to underadjust, that is, it insufficiently controls for initial differences. They argued that this was especially damaging for compensatory programs. For a compensatory program, the difference on the covariate favors the advantaged group. If covariance analysis underadjusts, then it would tend to make compensatory programs look bad as it did in the previous example. A program that had no effect would be made to look harmful and the effects of a slightly beneficial program might be wiped out. Cronbach and Furby (1970), however, state that the effect of misapplying covariance analysis is not so certain. In some cases it will underadjust and in other cases it will overadjust. Cronbach, Rogosa, Floden, and Price (1978) give an explicit mathematical formalization of the cases in which analysis of

Table 11.3. Estimates in Dollars of the Effects of the Manpower Training Program

Test	Index of Response	"Treatment Effect"
Analysis of covariance	.872	−$276
Covariance with reliability correction	.969	−$169
Raw change score analysis	1.000	−$134
Standardized change score analysis	1.084	−$40

covariance will under- and overadjust, but the mathematics of this very important paper go well beyond this text.

The nonequivalent control group design does not imply a single method of data analysis. Campbell and Stanley (1963) tend to give the mistaken impression that there is an ideal statistical analysis for each experimental and quasi-experimental design. It is true that the design can introduce certain specifications, but I can think of no design that can *always* be analyzed by a single statistical method.

The validity of each of the statistical methods discussed in this chapter depends on the causal model it presumes. One has no guarantee that any particular model is true. It is almost certainly the case that none of the models is exactly true. Special care must be given to various issues: unreliability in the covariate, growth rates over time, stationarity, and most importantly the process of selection.

12

Cross-lagged Panel Correlation

The models discussed in this chapter were developed exclusively for longitudinal data. Whereas the models discussed in the previous chapters that used either multiple regression or factor analysis to estimate structural parameters can be used in a longitudinal context, they were primarily developed to analyze cross-sectional data. Cross-lagged panel correlation (CLPC) was first suggested by Campbell (1963) and has been extensively reviewed by Kenny (1973, 1975*b*). In its simplest form CLPC involves two constructs, say X and Y, measured at two points in time, say 1 and 2. There are then four variables X_1, X_2, Y_1, and Y_2 and six correlations: two *synchronous correlations*, $\rho_{X_1Y_1}$ and $\rho_{X_2Y_2}$, two *autocorrelations*, $\rho_{X_1X_2}$ and $\rho_{Y_1Y_2}$, and two *cross-lagged correlations*, $\rho_{X_1Y_2}$ and $\rho_{X_2Y_1}$. As the name suggests, CLPC is the comparison of the cross-lagged correlations which can be expressed as a *cross-lagged differential*: $\rho_{X_1Y_2}$ minus $\rho_{X_2Y_1}$. Campbell's original suggestion is that if X caused Y, then the cross-lagged differential would be positive and if Y caused X the differential would be negative. (Unless otherwise stated, all correlations here are assumed to be positive.) Campbell and his students (Kenny, 1973, 1975*b*; Rickard, 1972; Rozelle & Campbell 1969) have elaborated the method. Another tradition (Bohrnstedt, 1969; Duncan, 1969; Goldberger, 1971; Heise, 1970; Pelz & Andrews, 1964) has suggested replacing CLPC with multiple regression or partial correlation analysis. Still another approach to panel data has been the application of factor analysis (Jöreskog & Sörbom, 1976). This approach usually involves oblique solutions in which factors are time specific. The fourth and oldest formal approach to panel data is the sixteenfold table described by Lazarsfeld (1972; Yee & Gage, 1968). This approach involves nominal data, usually dichotomous. Goodman (1973) has applied log-linear analysis to the sixteenfold table.

In Figure 12.1 there are three highly atypical but very interesting examples of CLPC. As someone who attended college in the late 1960s, I chose as examples marijuana use, women's consciousness, and Vietnam protest. The first example is taken from Jessor and Jessor (1977) and the two variables are marijuana behavior involvement and attitudinal tolerance of deviance. The marijuana variable is a four-item scale of the amount of marijuana usage and the attitudinal tolerance of deviance is a 26-item measure of the tolerance of "conventional" deviant behaviors like lying, stealing, and aggression. The sample consists of 188 males of high school age and the lag between waves is one year.

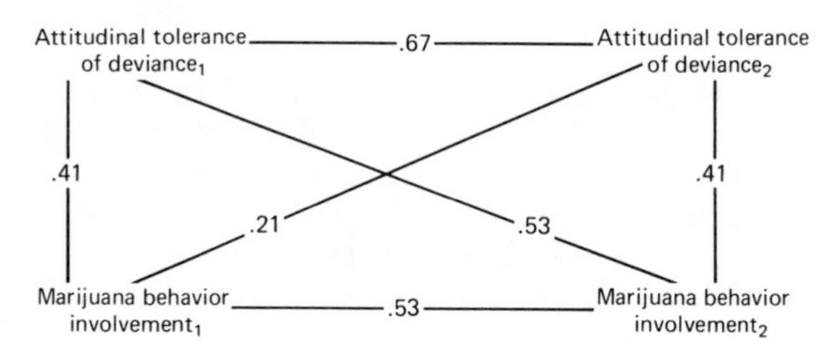

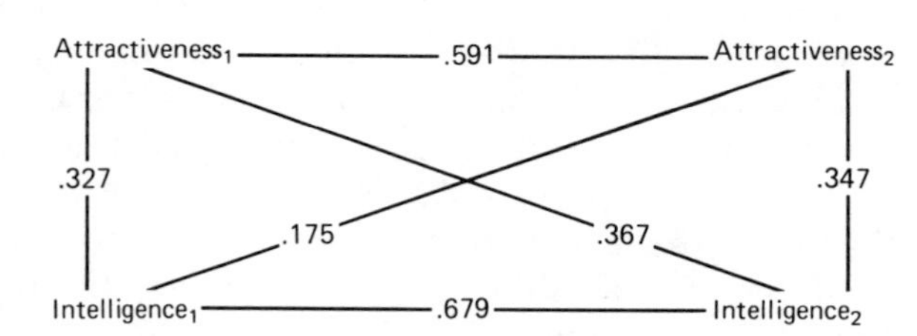

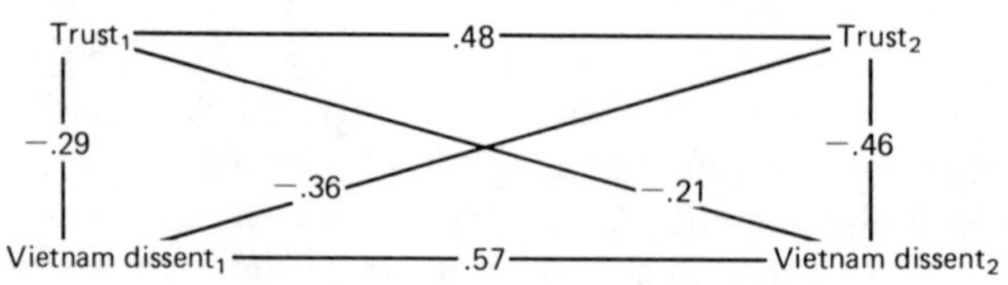

Figure 12.1 Crosslag examples.

The crosslags in Figure 12.1 suggest tolerance for deviance causes an increase in marijuana use.

The second example is taken from unpublished data of Z. Rubin. The subjects are 115 college-age female members of dating couples. The two measures are the woman's perception of her physical attractiveness and her perception of her intelligence. The lag between measurements is 15 months. The crosslags indicate that the opposite of the "dumb blond" hypothesis occurs. Instead they point to perceived attractiveness causing an increase in perceived intelligence.

The final example is taken from Bachman and Jennings (1975). The two variables are trust in the government and Vietnam dissent. The lag is one year and the sample consists of 1406 male late-adolescents. The crosslags indicate that dissent against the war in Vietnam caused a decrease in trust. Perhaps the lack of trust brought about by Vietnam laid the groundwork for public acceptance of the crimes of Watergate.

LOGIC OF CLPC

CLPC is a quasi-experimental design (Campbell & Stanley, 1963). At the heart of quasi-experimental inference is the attempt to rule out plausible alternative explanations of a causal effect, that is, biases or artifacts. In correlational analysis the chief alternative explanation of any causal effect is spuriousness. Any statistical relationship, be it simple correlation, partial correlation, or regression coefficient, can be attributed not to causality but to spuriousness. As discussed in Chapter 1, Suppes (1970) has even defined a causal relationship negatively as a nonspurious relationship. Ideally these spurious causes should be measured and controlled in the nonexperimental case.

True experiments control for spuriousness by random assignment to treatment conditions. As reviewed in Chapter 10, random assignment guarantees that there is no systematic relationship in the population between the treatment and the dependent variable given the null hypothesis of no treatment effect. Thus, any relationship between the treatment and the dependent variable that cannot be plausibly explained by chance is attributed to the causal effects of the treatment. Although random assignment permits researchers to make strong causal inferences, it brings with it some potentially burdensome methodological limitations. True experimentation rules out of consideration as independent variables any variable that cannot be manipulated and then randomly assigned. Many important variables, usually individual differences, are not subject to experimental manipulation as simply as the intensity of light. Returning to the examples in Figure 12.1, it is clearly difficult to manipulate variables like attitudinal tolerance of deviance, perceived intelligence, or Vietnam dissent. Researchers spend considerable time theorizing about intelligence, attitude change, extroversion–introversion, and evoked potential, but since these variables are attached to rather than assigned to the organism, they are studied more often as dependent rather than independent variables. To some degree the traditional stimulus–response or input–output orientation within psychology may reflect the limitation that experimental treatments be manipulatable variables. The requirement of manipulating the independent variable also prevents researchers from examining certain variables because of ethical considerations. For instance, malnutrition has been proposed as an important cause of children's cognitive ability, but it would be highly unethical to randomly assign children to levels of malnutrition. Thus, for practical and ethical reasons it is not always possible to use random assignment to control for spuriousness.

The null hypothesis of CLPC tested by equality of the crosslags is that the relationship between X and Y is due to an unmeasured third variable and not causation. Before causal models are entertained, the third variable explanation should be ruled out. The logic of true experimentation is similar. Before accepting that the treatment has an effect, the null hypothesis of sampling error must be ruled out. Given the inapplicability of true experimentation in numerous areas, CLPC can be used to test for spuriousness.

The null model for CLPC is illustrated in Figure 12.2. A third variable, Z_1, causes X_1 and Y_1 simultaneously. (Actually Z may cause X and Y with a lag, and the lag would be the same for both X and Y.) Over time Z changes and at time 2, Z_2 causes X_2 and Y_2. Given a model of spuriousness in Figure 12.2 the structural equations for X and Y are as follows:

$$X_1 = a_1 Z_1 + b_1 U_1 \quad [12.1]$$

$$X_2 = a_2 Z_2 + b_2 U_2 \quad [12.2]$$

$$Y_1 = c_1 Z_1 + d_1 V_1 \quad [12.3]$$

$$Y_2 = c_2 Z_2 + d_2 V_2 \quad [12.4]$$

where U, V, and Z are all uncorrelated with each other but are autocorrelated. Each unmeasured variable takes on a different role. Z is the unmeasured variable that brings about the relationship between X and Y and is called the *third variable*. The variable U includes all causes of

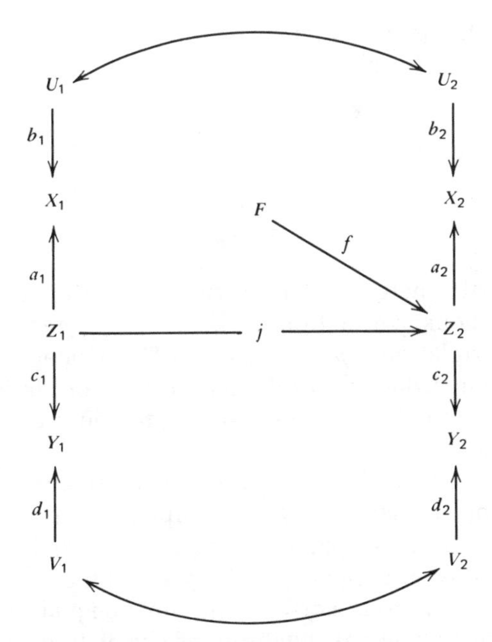

Figure 12.2 The null model for CLPc.

X besides Z. It includes true causes as well as errors of measurement. V plays the same role for Y. It is assumed that U and V are correlated over time. Although it is not necessary for much of what is to follow, it is assumed that Z is first-order autoregressive (see Chapter 9). The model in Figure 12.2 is clearly underidentified. There are seven free parameters and only six correlations. Interestingly enough, the autocorrelation of Z or $\rho_{Z_1Z_2}$ is identified by

$$\left(\frac{\rho_{X_1Y_2}\rho_{X_2Y_1}}{\rho_{X_1Y_1}\rho_{X_2Y_2}}\right)^{1/2}$$

The preceding value should be between plus or minus one since it is a correlation.

Since none of the other parameters are identified, how can spuriousness be tested? The key assumption of CLPC is *stationarity*. This assumption capitalizes on the fact that the same variables are measured at each point in time. By stationarity it is meant that a variable's causal structure does not change over time; that is, its structural equation is the same at both points in time. It is important to distinguish stationarity from *stability*. Stability refers to unchanging levels of a variable over time, whereas stationarity refers to an unchanging causal structure. For the two-wave, two-variable case stationarity implies

$$a_1 = a_2$$

and

$$c_1 = c_2$$

Given stationarity there are two overidentifying restrictions: equality of the synchronous correlations, $\rho_{X_1Y_1} = \rho_{X_2Y_2}$, and equality of the crosslagged correlations, $\rho_{X_1Y_2} = \rho_{X_2Y_1}$. The strategy of CLPC is to examine the synchronous correlations to test for stationarity, and if stationarity is satisfied the cross-lagged correlations can be used to test for spuriousness.

Unfortunately, even if the synchronous correlations are equal, stationarity may not be satisfied. For example, if $a_1/c_2 = a_2/c_1$, the synchronous correlations are equal whereas the crosslags are unequal. The opposite can also happen: if $a_1/c_1 = a_2/c_2$, the crosslags are equal whereas the synchronous correlations are unequal. Thus given the model in Figure 12.2, equal synchronous correlations is neither a necessary nor sufficient condition for equal crosslags. Quite obviously, it is highly questionable whether crosslags can be used to test for spuriousness in the simple two-wave, two-variable case: First, equal synchronous correlations are needed to demonstrate stationarity. Although it is rather common for the synchronous correlations to be equal, a fair percentage of the time they are not. Second, even if the synchronous correlations are equal, stationarity may still not hold. In such a case unequal crosslags do not indicate a violation of spuriousness but unstationarity.

Besides stationarity, the second key assumption of CLPC is *synchronicity*: Both variables must be measured at the same point in time. To see why this assumption is important, examine the model:

$$X_t = aZ_t + bU_t$$

$$Y_t = cZ_t + dV_t$$

$$Z_t = jZ_{t-1} + fF_t$$

where U, V, and Z are autocorrelated but not cross-correlated. The synchronous correlation is then ac and the crosslag of $\rho_{X_tY_{t+k}}$ is $acj^{|k|}$. Note that the synchronous correlation is a special case of the crosslag formula where $k = 0$. Now if X is measured at times 1 and 3 and Y at times 2 and 4, but if X_1 and Y_2 are considered wave 1 and X_3 and Y_4 are considered wave 2, then the "crosslags" would not be equal since

$$\rho_{X_1Y_4} = acj^3$$

and

$$\rho_{X_3Y_2} = acj$$

These unequal crosslags would not be due to a violation of spuriousness or even stationarity but due to a violation of synchronicity. Not surprisingly, the variables measured closer together in time correlate higher than those measured further apart in time. Synchronicity is then an important assumption of CLPC.

It is instructive to examine the formula for the crosslag in more detail: $acj^{|k|}$. Note that if j is positive, as it would be expected to be, the correlation between the two variables reaches an absolute maximum when the lag is zero. As the lag (i.e., k) increases, the correlation decreases in absolute value. Campbell has called this phenomenon *temporal attenuation* (Rozelle & Campbell, 1969) or *temporal erosion* (Campbell, 1971). Temporal erosion is an important empirical fact to remember in examining longitudinal data. The intuition behind temporal erosion is that two variables measured closer together in time should, *ceteris paribus,* be more highly correlated. This "fact" may not hold when either spuriousness or stationarity do not obtain. However, this is the very reason that temporal erosion has been emphasized by Campbell. Relationships that do not erode over time may indicate nonspurious effects. Note that given spuriousness and stationarity the crosslags must be less than or equal to in absolute value the synchronous correlations (Cook & Campbell, 1976). What governs this decrease in correlation is the rate of change of the third variable, Z.

At first glance synchronicity would seem to be an easy assumption to satisfy. Panel studies are defined as the replication at two different points in time of a cross-sectional survey on the same set of persons. However, because of the problems of *retrospection* and *aggregation,* synchronicity may not be so easily satisfied. Some variables in panel studies ask subjects to recall behaviors, attitudes, or experiences of the past. These questions either directly or indirectly ask subjects to retrospect. In some sense the data may not be generated at the time of measurement but at some time prior to measurement.

Another problem for synchronicity is aggregation. Many variables are aggregated or averaged over time. A good example of a measure of this type is grade point average. If grade point average is to be taken as a measure of ability, at what time point does it measure ability? It is actually an aggregation of performances evaluated by a teacher. The aggregation problem is well known in the econometrics literature where many of the important economic variables are aggregated across time, for example, gross national product and unemployment rates.

Thus both a lack of synchronicity and a lack of stationarity are potential explanations of a difference between cross-lagged correlations. If the model in Figure 12.2 is correct, then both stationarity and synchronicity together would imply equal crosslags. The null hypothesis that the crosslagged differential is zero is then a test of spuriousness.

What if the crosslagged differential is nonzero? Asymmetrical crosslags may indicate a causal effect; more generally they indicate that there is a factor that causes one of the measured variables and then causes the other measured variable at a later point in time. This factor, called the *causal factor*, is one of many factors that make up the causal variable. Saying "X causes Y" is shorthand for "something in X later causes Y." It need not be the case that the measure of X is valid or that the causal factor is the same as the true score of X. Although this problem of verbally explaining a causal effect is also present in true experiments, it is not as severe. One knows from an experiment that X causes Y but the experiment does not necessarily tell what in X causes Y. Different theoretical perspectives focus on different aspects of an experimental treatment to explain the same causal effect. The problem of interpreting cross-lagged differences centers on the *construct validity* of measures just as it does in experimental research. The more valid, reliable, and unidimensional the measure, the more straightforward is the interpretation.

To illustrate the difficulties in interpreting cross-lagged differences, consider the marijuana example in Figure 12.1. A naive interpretation would say that tolerance of deviance causes marijuana use. A more precise interpretation would be that some variable that causes tolerance for deviance later causes marijuana use. Perhaps there is a developmental sequence in which there is first tolerance for deviance and then marijuana use. Although lightning may precede thunder with a time lag, one would not argue that lightning causes thunder.

An alternative model to that in Figure 12.2 has S cause X simultaneously and Y with a lag of one unit of time. (Alternatively, S causes X with a lag of k units and Y with a lag of $k + 1$ units.) The equations then are

$$X_t = a_t Z_t + g_t S_t + b_t U_t$$

$$Y_t = c_t Z_t + h_t S_{t-1} + d_t V_t$$

In general the correlation between X and Y is

$$\rho_{X_tY_{t+k}} = a_t c_{t+k} \rho_{Z_tZ_{t+k}} + g_t h_{t+k} \rho_{S_tS_{t+k-1}}$$

For the special case in which there are two waves at times 1 and 2, the synchronous correlations are

$$\rho_{X_1Y_1} = a_1 c_1 + g_1 h_1 \rho_{S_0S_1}$$

$$\rho_{X_2Y_2} = a_2 c_2 + g_2 h_2 \rho_{S_1S_2}$$

and the crosslags are

$$\rho_{X_1Y_2} = a_1 c_2 \rho_{Z_1Z_2} + g_1 h_2$$

$$\rho_{X_2Y_1} = a_2 c_1 \rho_{Z_1Z_2} + g_2 h_1 \rho_{S_0S_2}$$

If stationarity is assumed, $a_1 = a_2$, $c_1 = c_2$, $g_1 = g_2$, and $h_1 = h_2$, then except for trivial cases the crosslags must be unequal. Note that this more general model diverges from the result of the model in Figure 12.2. The difference between crosslags assuming stationarity is

$$(1 - \rho_{S_0S_2})gh \qquad [12.5]$$

If it is now assumed that S behaves with a stationary first-order autoregressive process, it then follows that the synchronous correlations are equal and that it is possible for a crosslag to be larger than a synchronous correlation in absolute value. Such a result is not possible for the model in Figure 12.2.

Rozelle and Campbell (1969) and Yee and Gage (1968) have pointed out a difficulty in interpreting cross-lagged differences—competing, confounded pairs of hypotheses. There are two *sources* of a causal effect, X and Y, and two *directions* of that effect, positive and negative, making a total of four possible hypotheses. In terms of the previously elaborated causal model the source refers to which variable S causes first and the direction refers to whether g and h have the same or different sign. Finding $\rho_{X_1Y_2} > \rho_{X_2Y_1}$ is consistent with both X causing an increase in Y and Y causing a decrease in X. Finding $\rho_{X_1Y_2} < \rho_{X_2Y_1}$ is consistent with both Y causing an increase in X and X causing a decrease in Y. To illustrate the difficulties of confounded hypotheses consider the example taken from Kidder, Kidder, and Snyderman (1974) in Figure 12.3. The variables are the number of burglaries and the number of police for 724 United States cities. At first glance it appears that burglaries cause an increase in the number of police. An

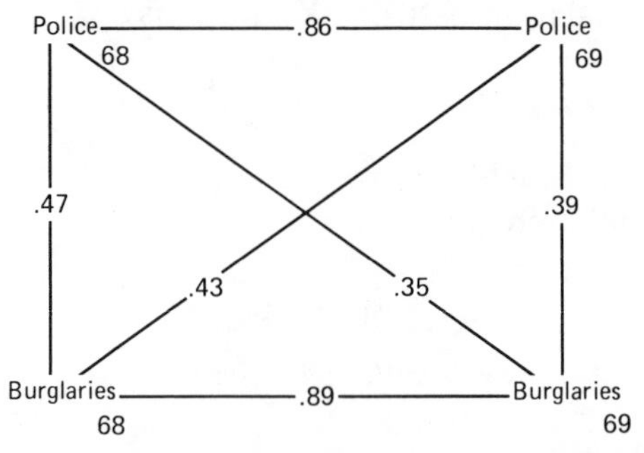

Figure 12.3 Confounded hypotheses: correlation of number of police and number of burglaries per capita measure in 1968 and 1969 in 724 cities.

alternative "law and order" explanation is that the number of police causes a *decrease* in the number of burglaries. Both hypotheses are equally plausible. The data are not consistent with two other hypotheses: police increase burglaries or burglaries decrease the number of police. These later hypotheses are not ruled out but their effects, if they exist, are overwhelmed by the effects of one or both of the two former hypotheses.

Rozelle and Campbell suggested a *no-cause baseline* be computed to test both of the confounded hypotheses. Their procedure is as follows:

a. Compute the test–retest correlations of both variables and correct them for attenuation.

b. Average these two correlations to obtain a measure of the stability.

c. Multiply the estimate by the average of the two synchronous correlations.

d. The resulting value is a no-cause baseline to which both crosslags can be compared.

The implicit logic of the no-cause baseline is that given temporal erosion, the crosslags should be less than the synchronous correlations by some factor. Given stationarity and spuriousness, the crosslags should equal the synchronous correlation times the autocorrelation of Z. Rozelle and Campbell assume that the autocorrelation can be estimated from the autocorrelations of X and Y. Unfortunately there are two difficulties with the Rozelle and Campbell baseline. First, it requires that the researcher have estimates of each variable's reliability because each autocorrelation must first be corrected for attenuation. Second, and more problematic, is the hidden assumption that all the nonerrorful causes of X and Y change at the same rate over time, that is, have the same autocorrelation. This assumption is called *homogeneous stability*. Evidence consistent with this assumption is that the two unattenuated autocorrelations are equal. Given the necessity of reliability estimates and homogeneous stability assumption, it would seem that the Rozelle and Campbell baseline is of limited practical use for longitudinal studies.

Although the sign of the synchronous correlations is neither a necessary nor sufficient condition for the direction of the causal effect, it is, nonetheless, suggestive of its direction. If the synchronous correlations are positive, they are supportive of X causing increases in Y or Y causing increases in X. Negative synchronous correlations indicate decreases. Moreover, occasionally the researcher may know the source of causation and the only empirical issue is the direction, or the direction is known and the only empirical issue is the source. In this way some of the confounded hypotheses can be ruled out a priori.

Given homogeneous stability the crosslags should always be smaller in absolute value than the synchronous correlations given spuriousness, stationarity, and synchronicity. Thus a crosslag larger than the synchronous correlations (assumed to be equal given stationarity) is indicative of a causal effect. It should be made clear that if X causes increases in Y and homogeneous stability is the case, then the crosslag from X to Y need not *necessarily* be larger than the synchronous correlations since both instability of spurious causes and misspecified causal lag would tend to make the crosslag smaller than the synchronous correlations.

INSIGNIFICANT DIFFERENCES

What does an insignificant difference between the cross-lagged correlations indicate? Strictly speaking one should not accept the null hypothesis of spuriousness, that is, the hypothesis that the variables do not cause each other but are cosymptoms of some set of common causes. There are several alternative explanations. First, it may be that both X and Y equally cause each other in a positive feedback loop making the crosslags equal. Without a no-cause baseline such a model cannot be distinguished from spuriousness. Second, it may be that X causes Y or vice versa, but the magnitude of the effect is too small to be detected. In my experience, it is very difficult to obtain statistically significant differences between cross-lagged correlations even when the sample size is moderate, 75 to 300. The cross-lagged differential depends on the stability of the causal factor, $\rho_{S_0S_2}$ from Equation 12.5. The more stable this factor is the smaller the differential. In the limiting case in which the factor does not change at all, the differential is zero. Cross-lagged analysis is, therefore, inappropriate for examining the causal effect of variables that do not change over time. For these variables their effects might best be diagnosed using other quasi-experimental models (see Chapter 11). (These models actually identify causal effects through unstationarity, that is, as increases in synchronous correlations.) Large cross-lagged differences are also difficult to obtain because the measured lag may not correspond to the causal lag. Normally the lag between measurements is chosen because of convenience, not theory, since theory rarely specifies the exact length of the causal lag. Moreover, for most social science theories it is doubtful that there is a single lag interval. Rather, the lag is variable across time or, as the econometricians call it, a *distributed lag*. Finally CLPC is most appropriate for the analysis of variables that have moderate or large correlations. My own experience is that the analysis of variables with low correlations (less than .3), yields disappointing and confusing results. In sum, CLPC requires at least moderate sample sizes, variables that change, lagged effects, and at least moderate synchronous correlations. To obtain such moderate correlations the instruments must be reliable. Never attempt a crosslagged analysis on measures with doubtful reliability.

Given the low power of CLPC the researcher should design the longitudinal study to include many replications. Ideally a cross-lagged difference should replicate across

a. Different time lags.
b. Different groups of subjects.
c. Different operationalizations of the same construct.

For instance, most of the causal effects in Crano, Kenny, and Campbell's (1972) study of intelligence and achievement can be summarized as abstract skills causing concrete skills. In one of the best empirical applications of cross-lagged analysis Calsyn (1973) demonstrates all three of the preceding types of replications to show that academic achievement causes academic self-concept.

SIGNIFICANCE TESTS

The hypotheses tested in a cross-lagged analysis are, first, the equality of synchronous correlations to test for stationarity and, second, the equality of crosslags to test for spuriousness. One cannot use Fisher's z transformation (McNemar, 1969, pp. 157–158) to test for the significance of the differences between these correlations since the correlations are correlated. One can, however, use a rather bulky but easily programmable test cited by Peters and Van Voorhis (1940) and attributed to Pearson and Filon.

Since the formula is not easily accessible it is reproduced here. Let 1, 2, 3, and 4 be variables, N be sample size, and

$$k = (r_{12} - r_{24}r_{14})(r_{34} - r_{24}r_{23}) + (r_{13} - r_{12}r_{23})(r_{24} - r_{12}r_{14}) + (r_{12} - r_{13}r_{23})(r_{34} - r_{13}r_{14}) + (r_{13} - r_{14}r_{34})(r_{24} - r_{34}r_{23})$$

The following then has approximately a standard normal distribution:

$$Z = \frac{(N)^{1/2}(r_{14} - r_{23})}{((1 - r_{14}^2)^2 + (1 - r_{23}^2)^2 - k)^{1/2}}$$

STATIONARITY REVISITED

The pivotal assumption on which cross-lagged analysis rests is stationarity. As demonstrated later, the assumption can be somewhat relaxed in the multivariate case; nonetheless it is still critical. Put in words, stationarity implies that a variable has the same proportions of "ingredients" at both points in time. More formally, it implies that the structural equations are the same. The exogenous variables may be changing, but they are weighted the same way at both times. Others have called this assumption a *steady-state process*.

Stationarity cannot be guaranteed by using the same operationalization at both points in time. It may well be that the same test measures different constructs at different times. Unfortunately stationarity may be least appropriate in the situation in which it is most needed: rapid developmental growth. Developmentalists today are emphasizing the discontinuities of growth. Implicit in such stage theories of growth is that the causal model shifts. Variables that are important at one stage of development are irrelevant at other stages of development. Instead of assuming stationarity, it may be more reasonable to assume that a variable becomes more causally effective over time. This effect would be indicated by not only asymmetrical crosslags, but also changing (usually increasing) synchronous correlations. Thus in many nontrivial instances, changing synchronous correlations may be indicative of causal effects.

One good rule of thumb is that stationarity is more plausible in cases in which the lag between measurements is short. For instance, I found in the analysis of four-wave data sets that the adjacent waves were relatively more stationary, than were the more distant waves.

BACKGROUND VARIABLES

Very often panel studies contain measures like sex, ethnicity, social class, and other background variables which are potential sources of spuriousness. There are two different strategies for handling these background or social grouping variables. The first is to perform separate analyses on each sex, race, or social group. The second is to subtract out the effects of the background variable. The first strategy is preferred if different causal patterns are expected for different social groups. For instance, Crano, Kenny, and Campbell (1972) found contrasting causal relationships for lower and middle class children in the relationship between intelligence and achievement. However, sample size often prohibits this strategy.

The second strategy—subtracting out the effects of background variables—can be done by computing partial correlations between the relevant variables controlling for the background variables. If the background variables are nominally coded then dummy variables can be created for them. This procedure assumes that the causal processes are the same within social groups although the groups may differ in mean level. Controlling for background variables often increases the stationarity of the data. After the background variables are partialled, the synchronous and cross-lagged *partial* correlations can be examined.

There are two helpful rules in choosing variables to partial out. First, the variable to partial out should independently explain, at least, a moderate amount of variance. Otherwise nothing is changed by the partialling. Second, D. T. Campbell has suggested that any control variable should in principle be able to explain as much variance of the time 1 variables as the time 2 variables. For instance, imagine a study of cognitive skills that had only a time 2 measure of intelligence. Given temporal erosion, the intelligence measure will correlate more highly with the time 2 measures than the time 1 measures, and, therefore, it would be inappropriate as a variable to partial on.

THREE-WAVE, TWO-VARIABLE MODEL

The model in Figure 12.2 can be extended to three waves. The general equations are

$$X_t = a_t Z_t + b_t U_t$$

$$Y_t = c_t Z_t + d_t V_t$$

The usual assumptions are made that Z, U, and V are autocorrelated but not cross-correlated. If Z is first-order autoregressive, then $\rho_{13} = \rho_{12}\rho_{23}$ for Z. It then follows that the following two vanishing tetrads hold:

$$\rho_{X_1Y_2}\rho_{X_2Y_3} - \rho_{X_1Y_3}\rho_{X_2Y_2} = 0$$

$$\rho_{X_2Y_1}\rho_{X_3Y_2} - \rho_{X_3Y_1}\rho_{X_2Y_2} = 0$$

Note that both of these tetrads hold even if there is not stationarity: $a_1 \neq a_2 \neq a_3$ and $c_1 \neq c_2 \neq c_3$. Even if Z is not autoregressive, the following overidentifying restriction holds:

$$\rho_{X_1Y_2}\rho_{X_2Y_3}\rho_{X_3Y_1} - \rho_{X_2Y_1}\rho_{X_3Y_2}\rho_{X_1Y_3} = 0$$

Unfortunately these overidentifying restrictions are of little practical use. In the first place, three-wave data are difficult to obtain. Second, it must be presumed that the causal lag is or is very near to the measured lag between the two waves. If the lag is much different, the tetrads still vanish. Finally the tetrads presume a first-order autoregressive process, and such an assumption may often be unrealistic.

For example, examine the crosslags in Figure 12.4. An example of three waves from Jessor and Jessor (1977) is contained. The synchronous correlations are stationary, and the crosslags clearly indicate a causal effect from attitudinal tolerance of deviance (ATD) to marijuana behavior involvement (MBI). However, both the tetrads vanish: $\chi^2(1)$ equals 1.23 and .03.

Just as a two-wave study does not begin to solve all the inference problems of the cross-sectional study, the three-wave study is no panacea. Quite bluntly, the delivery of longitudinal studies falls far short of its inflated promises. The view that "one more wave" is all that is needed to make the conclusions of the study airtight is clearly mistaken. Longitudinal studies are very expensive and usually have severe attrition problems. They should only be undertaken with the sober realization of their limitations.

TWO-WAVE, N-VARIABLE MODEL

For the previous models of CLPC, it was assumed that only a single unmeasured exogenous variable caused both endogenous variables. Now it will be assumed that all the endogenous variables are caused by

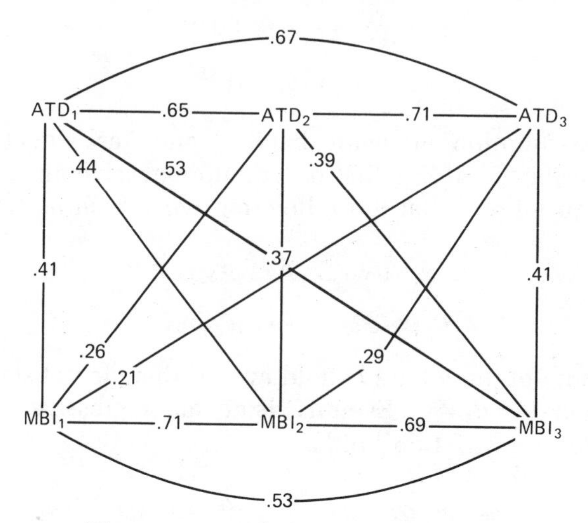

Figure 12.4 Three-wave example.

a set of p exogenous variables, $F_1, F_2, F_3, \ldots, F_p$. These p variables are ordinarily all unmeasured and may be considered factors. The equation for endogenous variable X_i at time 1 is

$$X_{i1} = \sum_{j=1}^{p} a_{ij} F_{1j} + c_{i1} U_{i1}$$

where a_{ij} are the common factor loadings and U_{i1} is a unique factor for X_{i1} uncorrelated with the Fs. The Us can be considered errors of measurement and may be correlated over time. All the factors and measured variables have unit variance. Similarly the equation for X_i at time 2 is

$$X_{i2} = \sum_{j=1}^{p} b_{ij} F_{2j} + c_{i2} U_{i2}$$

All the unmeasured variables are standardized at both time points. It is also assumed that all the common factors, the Fs, are orthogonal to each other. Although the orthogonality assumption is totally unrealistic, it greatly simplifies the algebra. Later in this chapter, the model is modified to allow the factors to be correlated.

The stationarity assumption is

$$k_i = \frac{b_{i1}}{a_{i1}} = \frac{b_{i2}}{a_{i2}} = \frac{b_{i3}}{a_{i3}} = \cdots = \frac{b_{ip}}{a_{ip}} \qquad [12.6]$$

This assumption has been called quasi-stationarity (Kenny, 1975b). It simply states that the path coefficients for each variable increase proportionately over time. The term k_i^2 has a simple interpretation. It is the ratio of the X_{i2} communality to X_{i1} communality. This is so since the communality of X_{i2} is $\sum_{j=1}^{p} b_{ij}^2$ and the communality of X_{i1} is $\sum_{j=1}^{p} a_{ij}^2$. But $\Sigma b_{ij}^2 = k_i^2 \, \Sigma a_{ij}^2$ given Equation 12.6. It then follows that the ratio of communalities is k_i^2. Since for a variable with no specific factor the communality and reliability are identical, one can view k_i^2 as reliability ratio assuming no specific variance. Quasi-stationarity also implies that the standardized structural equation of each variable is invariant with respect to time once the unique factor has been omitted from the equation and the coefficients are accordingly changed to standardize the equation.

As has been shown in Kenny (1973) these communality ratios can be estimated in a straightforward manner. The correlation of X_i with X_j at time 1 is $\sum_{k=1}^{p} a_{ik}a_{jk}$ and $\sum_{k=1}^{p} b_{ik}b_{jk}$ at time 2. But given quasi-stationarity, it then follows that $\Sigma b_{ik}b_{jk} = k_i k_j \Sigma a_{ik}a_{jk}$. Let q_{ij} stand for the ratio of the time 2 correlation of X_i with X_j to the time 1 correlation of the same variables, and then q_{ij} simply equals $k_i k_j$. Now given three variables X_i, X_j, and X_m it follows that

$$k_i^2 = \frac{q_{ij}q_{im}}{q_{jm}} \qquad [12.7]$$

Both k_j^2 and k_m^2 can be obtained in a similar fashion:

$$k_j^2 = \frac{q_{ij}q_{jm}}{q_{im}}$$

$$k_m^2 = \frac{q_{im}q_{jm}}{q_{ij}}$$

If the communality ratio k_i^2 is larger than one, the communality is increasing over time; if less than one, the communality is decreasing; and if one, the communality is unchanged. Although it is theoretically possible for k_i to be negative, it is unlikely. It would be indicated by changes in sign of the synchronous correlations and probably negative autocorrelations. Since k_i is not likely to be negative, one can routinely take the positive root of k_i^2.

It follows from the earlier equations for the synchronous correlations that if the time 1 synchronous correlation of X_i and X_j is multiplied by

$(k_i k_j)^{1/2}$ and the time 2 correlation is divided by the same value, the two synchronous correlations should be equal. With this correction, both correlations should now equal $(k_i k_j)^{1/2}\Sigma a_{ik}a_{jk}$.

In a similar fashion the cross-lagged correlations can be corrected to be made equal. The cross-lagged correlations between X_i and X_j are $\sum_{k=1}^{p} a_{ik}b_{jk}\rho_k$ where X_i is measured before X_j and $\sum_{k=1}^{p} b_{ik}a_{jk}\rho_k$ where X_j is measured before X_i and ρ_k is the autocorrelation of F_k at time 1 and F_k at time 2. Given quasi-stationarity one crosslag is $k_j\Sigma a_{ik}a_{jk}\rho_k$ and the other is $k_i\Sigma a_{ik}a_{jk}\rho_k$. If the first crosslag is multiplied by $(k_i/k_j)^{1/2}$ and the second by $(k_j/k_i)^{1/2}$, both corrected crosslags should be equal. Unequal corrected crosslags would then indicate a violation of spuriousness.

The very same conclusions can be obtained for an oblique set of factors if stationarity assumptions are imposed. It must be assumed that the synchronous correlation of F_i with F_j does not change over time and that the crosslags of F_i and F_j are equal; that is, $\rho_{i_1j_2} = \rho_{i_2j_1}$.

SIGNIFICANCE TESTING AND ESTIMATION

An example is perhaps the best way to illustrate the rather detailed process of estimation. The example is taken from the Educational Testing Service (ETS) evaluation of *Sesame Street* by Ball and Bogatz (1970). The data to be discussed here were obtained from the Cook, Appleton, Conner, Shaffer, Tamkin, and Weber (1975) reanalysis of the data. The sample to be discussed includes 348 preschoolers from five cities. The children are predominately disadvantaged. All the children were encouraged to view *Sesame Street* by the ETS staff. The children were measured at two points in time with a six-month lag. During six months the children watched, to differing degrees, *Sesame Street*.

The four dependent variables to be considered are Body Parts, a 32-item test to measure knowledge of body parts; Forms, an 8-item test to measure recognition of geometric figures; Letters, a 65-item test to measure recognition of letters of the alphabet; and Numbers, a 54-item test to measure number recognition. All tests were devised by ETS and were tailored to the objectives of *Sesame Street*.

In Table 12.1 above the diagonal is the partial correlation matrix for the four measures at the two points in time. For this analysis 10 variables were controlled or partialled out making the effective sample size 338. Some of these are dummy variables to control for sex, city, and race, and others are interval measured variables like age and social class.

Table 12.1. ***Sesame Street*** **Correlations with Background Variables Partialled Out**[a]

	Time 1				Time 2			
	Body Parts	**Forms**	**Letters**	**Numbers**	**Body Parts**	**Forms**	**Letters**	**Numbers**
Time 1								
Body Parts	1.000	.469	.248	.491	.434	.287	.241	.343
Forms		1.000	.337	.515	.304	.292	.319	.380
Letters			1.000	.570	.167	.203	.350	.343
Numbers				1.000	.290	.344	.448	.571
Time 2								
Body Parts					1.000	.497	.425	.506
Forms						1.000	.566	.595
Letters							1.000	.745
Numbers								1.000

[a]N = 338.

To employ the methods elaborated in Kenny (1975*b*), first the earlier described q_{ij} matrix is created by dividing the time 2 synchronous correlation by the time 1 synchronous correlation between the same variables. For instance, the q for letters and forms is .566/.377 = 1.680. Table 12.2 presents the q matrix. One can solve for the communality ratios by substituting the sample based estimates into Equation 12.7. There are three solutions for each of the four communality ratios. In general given n variables there are $(n-2)(n-1)/2$ solutions. The solutions are

Body Parts:	1.081,	.945,	1.351
Forms:	1.039,	1.188,	1.485
Letters:	2.716,	2.173,	1.900
Numbers:	1.124,	.786,	.899

In theory all the preceding estimates should be the same, but they differ because of sampling error.

The problem now is how to pool the three estimates of the communality ratios. Kenny (1975*b*) suggests the following procedure. The three estimates for Body Parts are

$$\frac{(.497)(.425)(.337)}{(.469)(.248)(.566)}, \frac{(.497)(.506)(.515)}{(.469)(.491)(.595)}, \frac{(.425)(.506)(.570)}{(.248)(.491)(.745)}$$

Each estimate is the product of three correlations divided by three correlations. What can then be done is simply to sum the numerators of each estimate and sum the denominators. Then divide the two in order to obtain an estimate of the communality ratio. For instance, for Body Parts the pooled estimate is

$$\frac{(.497)(.425)(.337) + (.497)(.506)(.515) + (.425)(.506)(.570)}{(.469)(.248)(.566) + (.469)(.491)(.595) + (.248)(.491)(.745)} = 1.101$$

Table 12.2. The Time 2 Synchronous Correlation Divided by the Time 1 Synchronous Correlation

	Body Parts	Forms	Letters
Forms	1.060		
Letters	1.714	1.680	
Numbers	1.031	1.155	1.307

The remaining pooled communality ratios are 1.277 for Forms, 2.135 for Letters, and .935 for Numbers. Note that all the estimates are compromises of the three unpooled estimates. A problem does arise if some of the correlations are negative. In such cases the absolute values should be summed.

Next it must be determined whether the synchronous correlations are equal after they have been corrected for shifts in communality. The hypothesis to be tested is

$$\rho_{i_1j_1}(k_ik_j)^{1/2} = \frac{\rho_{i_2j_2}}{(k_ik_j)^{1/2}}$$

The Pearson–Filon test can be used to compare the two correlated correlations. There are, however, two problems with this method. It ignores, first, that the ks are sample estimates and, second, that the ks were deliberately chosen to make the synchronous correlations equal.

In Table 12.3 there is the Pearson–Filon test of the difference between synchronous correlations both with and without correction. It shows that for three of the pairs of the synchronous correlations there is a significant difference. All three of these pairs involve the Letter variable, which is not surprising since that variable has the communality ratio that is most different from one. After correcting for shifts in communality, these three significant differences between the synchronous correlations disappear. It then appears that the correction is helping and quasi-stationarity is indicated.

In a similar fashion the difference between crosslags can be tested. Again there is the problem that it must be assumed that the ks are population values. In Table 12.4 there is the Pearson–Filon test of the

Table 12.3. Pearson–Filon Test of Equality of the Synchronous Correlations

	Uncorrected for Shifts in Communality	Corrected for Shifts in Communality
Body Parts–Forms	.501[a]	− .972
Body Parts–Letters	2.811	.566
Body Parts–Numbers	.292	.150
Forms–Letters	3.924	.127
Forms–Numbers	1.595	.614
Letters–Numbers	4.243	−1.231

[a]The value is positive if the time 2 synchronous correlation is larger than the time 1 correlation and negative if the time 1 correlation is larger.

Table 12.4. Pearson–Filon Test of Equality of Crosslags

	No Correction	Correction[a]
Body Parts–Forms	− .275[a]	− .628
Body Parts–Letters	1.107	.107
Body Parts–Numbers	.905	1.350
Forms–Letters	1.801	.770
Forms–Numbers	.625	1.611
Letters–Numbers	−2.015	1.095

[a]The value is positive if the first variable in the row "causes" the second, and negative if the second variable "causes" the first.
[b]Correction for shifts in communality.

difference between crosslags. There is one significant difference between crosslags before correction: with Letters and Numbers. Before the reader begins to concoct all sorts of explanations how the learning of numbers causes the learning of letters, note that after correction this difference disappears.

CONCLUSION

CLPC is a valuable technique for ruling out the plausible rival hypothesis of spuriousness. It should not be viewed as only an intuitive approach but as a formal method with assumptions. This chapter has inordinately emphasized alternative explanations of crosslag differences in order to present the reader with a list of problems much in the same way that Campbell and Stanley (1963) do for quasi-experimental designs.

CLPC is, however, largely an exploratory strategy of data analysis. My own suspicion is that its main use will be in uncovering simple causal relationships between uncontrolled variables. What would then follow is either the refinement of both the variables and the process in controlled settings or the estimation of causal parameters of the system by structural equation models. The career of a hypothesized causal relationship might be as follows: first, the consistent replication of a cross-sectional relationship; second, the finding of time-lagged relationships between cause and effect; third, the finding of cross-lagged differences; and fourth, an experiment in which the causal variable is manipulated. Obviously, these steps may often overlap, some may be omitted, and the order may be different. I hope to emphasize that CLPC plays only an intermediary role in social science, between the correlation and a well-elaborated structural model.

13

Loose Ends

In this final chapter both technical and conceptual issues concerning causal modeling are covered. Hopefully some closure is brought to what is certainly a limitless topic. The chapter is divided into two sections. The first section considers technical issues of linearity, ordinal and nominal variables, and standardization. The second section considers conceptual problems in model testing. Let me now try to tie together the loose ends that remain.

TECHNICAL ISSUES

Linearity

In this context linearity refers to a straight-line functional relationship between the cause and effect and not the level of measurement of either variable or the additivity of effects. These last two topics are discussed later in this chapter. The issue of linearity then concerns itself with whether there is straight-line relationship between X and Y. The simplest and most intuitive test of linearity is to examine the scatterplot between X and Y. One should always remember that behind every correlation lies a scatterplot; a correlation or regression coefficient is only a number that tries to capture the complex meaning of the scatterplot. While the old eyeball method can often be very useful in determining nonlinearity, in marginal cases your left eyeball may disagree with your right eyeball. You may then need a more formal way to assess the degree of nonlinearity.

One common method of testing for nonlinearity is polynomial regression. The effect variable, Y, is regressed on $X, X^2, X^3, \ldots, X^n$. If any variable besides X significantly predicts Y, then nonlinearity is indicated. Be aware of problems of large numbers (e.g., 25^7) and multicollinearity.

A well-known measure of relationship, both linear and nonlinear, is η^2, the correlation ratio. A second strategy for testing nonlinearity based on η^2 is to compute $\eta^2 - \rho^2$ which is a measure of the percentage of Y variance explained by nonlinearity. Operationally this becomes a bit computationally messy since η^2 can only be estimated when X is divided into discrete categories. If X is not naturally in such categories, it must be artificially divided into groups, for example, for intelligence 80–89, 90–99, 100–109, and so on. The estimate of η^2 is obtained from a one-way analysis of variance with the discrete levels of X as the independent variable and Y as the dependent variable. The value for $\hat{\eta}^2$ is then $SS_{BG}/(SS_{BG} + SS_{WG})$ where SS_{BG} is the sum of squares between groups and SS_{WG} the sum of squares within groups. Now $\hat{\rho}^2$ should not be computed (as is sometimes mistakenly done) by correlating X with Y. Rather the means of the Xs within categories are correlated with raw Y variable. It then follows that under the null hypothesis of $\eta^2 = \rho^2$

$$F_{k-2,N-k} = \frac{(\hat{\eta}^2 - \hat{\rho}^2)SS_{TOT}/(k - 2)}{MS_{WG}}$$

where k is the number of categories that were created. A more efficient method is analysis of covariance. The variable X is the covariate and discrete levels of X are the "treatments." If there is a main effect for treatments then nonlinearity is indicated. This method is more powerful than the analysis of variance method since the MS_{WG} is reduced due to the covariate.

The assumption of linearity should not be tested bivariately. It could very well be that the relationship between X and Y is nonlinear but once other variables are controlled the relationship becomes linear. Thus, it is more appropriate to examine the relationship between X and residuals of Y, if they can be computed.

Given that the relationship between X and Y is nonlinear, what is to be done? The usual strategy is a transformation. Either the exogenous or endogenous variable is transformed into a new metric. Following Kruskal (1968) it is useful to distinguish between one- and two-bend transformations. As suggested by the name, one-bend transformations are functions whose curves have a single bend, while two-bend transformations are curves with two bends. Power transformations of the form $X_t = aX^p$, where X_t is the transform of X, are called one-bend transformations. Transformations like arcsin, logit, probit, and Fisher's Z are two-bend transformations. One-bend transformations are generally more appropriate for variables that have a lower limit of zero and no upper limit. Two-bend transformations are useful for variables that have a lower and upper limit, for instance zero and one.

The three most common one-bend transformations are logarithm, square root, and reciprocal. The logarithm is most useful for amounts, for example, dollars. More generally, the log transformation is useful when it is suspected that two variables combine multiplicatively to cause the dependent variable. For instance, the Cobb–Douglas law states that capital and labor multiply to cause gross national product. The square root transformation is most useful for variables that are assumed to be distributed as poisson. Good candidates for a square root transformation, therefore, are counts, for example, the number of behaviors per unit time. The third one-bend transformation is the reciprocal or $1/X$. This transformation is especially useful for a variable like time to response or latency, where the transformed variable is now in the metric of speed. With any transformation it is important to be able to interpret the metric of the transformed variable.

The two-bend transformations are useful for variables that are proportions or could easily be turned into a proportion. For instance, the number of items passed in an intelligence test could be divided by the total number of items and would then become a proportion. The most radical transformation is the logit which is the natural logarithm of the odds. The odds of a proportion P is simply defined as $P/(1-P)$. Slightly less radical is probit. Probit, which is used by biologists, simply takes the probability and looks up the corresponding standard normal deviate (Z value) that would yield that probability. Since biologists do not seem to like negative numbers, it is customary to add five to this value. The least radical two-bend transformation is the arcsin, a transformation psychologists seem to prefer in analyzing proportions.

There is a final two-bend transformation that is commonly employed. It is Fisher's Z transformation which is a cousin of the logit transformation. This transformation is usually applied to correlations.

One obvious point should be made here. One-bend transformations should not be made on negative numbers, the log and reciprocal on zero, and logit and probit on zero or one.

Tukey (1977) outlines a number of useful methods by which the appropriate transformation can be chosen empirically. It is beyond the scope of this book to restate these methods. A useful point, though, is that when exploring which transformation to apply, one should work with medians since the median of transformed set of data is the transform of the median while this does not necessarily hold for the mean.

It has been assumed throughout that the causes of an endogenous variable add up. Of course, in some cases nonadditive models may be just as plausible if not more plausible. The whole may not be the sum of its parts. Within the experimental tradition there are methods to distinguish whether the independent variables add or multiple. Also nonlinear least squares (Draper & Smith, 1966) is a feasible computing method, but in the nonexperimental case the functional form in which the exogenous variables combine is best determined a priori and not in any simple a posteriori manner.

Ordinal and Nominal Variables

Just as every adolescent goes through an identity crisis, it seems that every researcher goes through a "measurement crisis." At some point he or she realizes that the interval assumptions of most social scientific measures are just assumptions. A little reflection then reveals that, at best, most measures are only ordinal. The researcher then vows to discard the correlation and the mean for the rank order coefficient and the median. But just like the crisis of an adolescent, it seems only to be important to have the crisis not to resolve it.

It is indeed true that there is a giant leap of faith in assuming that our measuring instruments are interval, but we are often forced to make just such an assumption. Unfortunately sometimes we are faced with a measure that is clearly ordinal. There is no meaningful rule for assigning numbers to objects. In such a case we might try to determine numbers by scaling methods, for example, multidimensional scaling methods.

If the exogenous variable is ordinal one can simply create a set of dummy variables as was done in Chapter 10. If both the endogenous and exogenous variables are ordinal, there is an interesting empirical scaling method that gives scale weights to the categories of both variables to maximize the correlation between them (Bonacich & Kirby, 1975).

In recent years there has been a great deal of statistical work on the analysis of nominal variables. The model that is analogous to the regression models in Chapter 4 is called *logit analysis* (Fienberg, 1977), and the model that is analogous to factor analysis is called *latent structure analysis* (Goodman, 1974). The two preceding sources should prove useful introductions to readers interested in either topic. To give the reader an example of the power of these methods consider the following example taken from Werts, Linn, and Jöreskog (1973). Given three measured dichotomies A, B, and C, let there be a dichotomy Z such that within levels of Z, variables A, B, and C are independent. This is a latent structure model similar to the single-factor models discussed in Chapter 7. Werts, Linn, and Jöreskog show that not only can the loadings be estimated but the marginals of the underlying dichotomy can be as well. One potential use of this technique might be in clinical assessment. For instance, say three clinicians *independently* made judgments about whether each member of the patient population was schizophrenic or not. It could then be determined what is the actual percentage of the schizophrenics. Also, the hit rate of each clinician can be measured, and from the hit rates one can determine the probability a patient is a schizophrenic given the judgments of the clinicians. Of course, independence of judgments is of the essence.

Interaction

Unfortunately, the topic of interaction has received scant treatment in the causal modeling literature. To some degree researchers do not search for interactions because they are afraid to find them. The approach here is that an interaction implies that a causal law of the form "X causes Y" must be modified to "X causes Y depending on Z." The variable Z is said to modify the X causes Y relationship. This notion of interaction as a modifier of a causal law has led some to suggest the following path symbols for interaction:

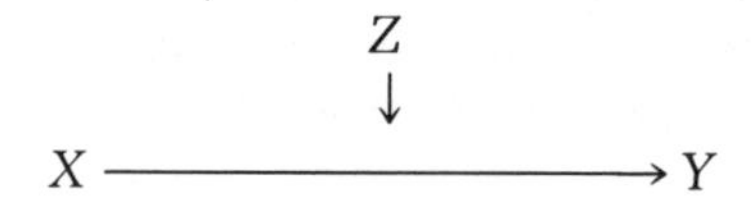

Note that the path from Z does not point to X or Y, but rather to the path or causal relationship between X and Y. This view of interaction is somewhat different from the usual analysis of variance view of interaction in which the two components of the interaction are equal partners in the relationship. The view here is more like the view of Campbell and Stanley (1963) who define an interaction as a threat to external validity. The choice of whether Z modifies XY, the causal relationship between X and Y, or X modifies ZY relationship is arbitrary. For instance, if age and motivation interact in causing athletic ability, a developmentalist would say motivation modifies the developmental age trends and a motivational psychologist would say age modifies the effect of motivation. Both would be correct, but both would be aided by the idea of modification.

The usual way that interaction is modeled is by the inclusion of multiplicative terms in the model. For instance for our X, Y, Z example we would have

$$Y = aX + bXZ \quad [13.1]$$

To capture the idea of modification we simply factor X out of 13.1 to obtain

$$Y = (a + bZ)X \quad [13.2]$$

Thus the structural coefficient for X is $a + bZ$. It is a variable since we do not know how much X affects Y without knowing Z. Take a simple case where Z can only take on the two values of 0 and 1 and $a + b = 0$. When Z is zero the coefficient is a, and when Z is 1 the coefficient is zero.

Researchers should be encouraged to introduce interaction terms. One should note that ordinarily the XZ and X will be highly correlated, depending largely on whether X has a nonzero mean. Thus, the researcher may have low power in testing interactions. Currently there is much interest in the addition of multiplicative terms to a linear model (Bohrnstedt & Marwell, 1977; Cohen, 1978).

As was just stated, it may be useful to include multiplicative terms in an equation to test for specification error. The presence of a significant interaction term might indicate either a need to transform, a nonadditive combination of variables, or even the omission of important third variables.

Standardization

This text has relied very heavily on the use of standardized coefficients. To many researchers who exclusively rely on correlation it may come as a surprise that America's best known statistician, John Tukey, has said:

> Sweeping things under the rug is the enemy of good data analysis. Often, using the correlation coefficient is "sweeping under the rug" with a vengeance. (1969, p. 89)

Others have called correlation and standardization the devil's tools (Birnbaum, 1973). It would then seem that standardization is highly controversial. Let us consider four criticisms of standardization.

1. Standardized measures are inappropriate for generalizing across populations.
2. The unstandardized metric is more interpretable.
3. Standardization will not allow certain specifications or at least make them clumsy.
4. Standardization introduces statistical complexities.

Consider the first point, that standardized coefficients are not as comparable across populations as unstandardized coefficients. It is more theoretically plausible that structural parameters may be invariant across different populations when they are in the unstandardized metric than in the standardized metric. Normally, invariance of the unstandardized metric does not imply invariance of the standardized metric. For instance, if the structural equations for two populations are identical, the standardized path coefficients will ordinarily differ if the variances of the endogenous variable differ. Thus, the well-known finding that beta weights and R^2 are usually larger for whites than blacks is often an artifact due ordinarily to the larger variance in whites than blacks. A good rule of thumb is that when comparing structural coefficients across populations, it is usually best to compare the unstandardized coefficients.

The preceding rule is easy to apply with causal models with measured variables. It is not as instructive for models with unobserved variables. Take, for instance, the simple case in which there is one factor that causes three measured variables with uncorrelated disturbances. We can take the two strategies. We could assume that the variance of the unmeasured variable is invariant across populations and let the "factor loadings" be variable, or we could assume that the loadings were invariant. If we take the second strategy we must still fix one factor loading. Suffice it to say that we cannot in this case nor in most others allow both the factor loadings and variances to vary across populations.

The second criticism of standardization is that the unstandardized metric is a more interpretable one. If the raw metric has some natural meaning, then all is lost by standardization. It is nice to know how many dollars of earned income is caused by each year of education, the number of items learned for each season of *Sesame Street* that is watched, and number of pecks for each reinforcement. However, interpretability is often a subjective issue. Sadly, the metrics in most social science studies are nearly meaningless. It may be the case that knowing that if a unit is one standard deviation above the mean on X, then the unit is b units of a standard deviation from the mean on Y is more meaningful. Perhaps the biggest challenge to social scientists is to make measures more naturally interpretable.

The third criticism of standardization is that it sometimes prevents the researcher from making specifications that allow for identification. Let us give an example. In Chapter 9 we discussed a model of Wiley and Wiley that stated that the error variance of a measure would remain stable over time while the true variance might change. If the researcher has standardized the variables and discarded the variances, it is impossible to make such a specification. If the variances are not lost it is still possible to standardize and then assume

$$\frac{1 - \rho_{X_1X_1}}{1 - \rho_{X_2X_2}} = \frac{V(X_2)}{V(X_1)}$$

It is always possible that one can standardize and estimate the parameters if the variances are saved. It is then critical always to compute and report variances, as well as means. Even if you are not interested in the means or variances, someone else may be.

It is also possible, though not common, that some specifications or overidentifications can only occur for the standardized metric. For instance, the fan spread hypothesis implies an overidentification that holds for treatment–effect correlations but not covariances.

In certain cases the unstandardized metric is much simpler. For instance, with standardization it is impossible to make the classical specification that a measured variable equals the true score plus errors of measurement. This created complications for the analysis in Chapter 5 on measurement error. Yet is was a simple matter to handle.

The final reason for unstandardized coefficients is that the standardized coefficients imply dividing by a sample estimate, the standard deviation. This should increase confidence intervals and alter tests of significance. Although these problems are beyond the scope of this text and the competence of its author, we should recognize that we have created a statistical problem by standardization. We leave it to the statisticians to develop solutions to these problems.

In general, unstandardized coefficients should be preferred. However, one should be flexible. Often times, I compute both. In Chapter 3 we gave a very simple formula to destandardize and another formula to standardize an unstandardized coefficient. This text has overly concentrated on the standardized metric. This was done for two reasons. First, most researchers are more familiar with the standardized mode. It seems to me simpler to motivate causal modeling by using a language and notation familiar to the audience. Second, in my opinion, and others differ here, the standardized metric is just much simpler to work with. One need not worry about normalization rules and the like. It seems to me that standardization is especially simpler when one takes a simple high school algebraic approach as was done in the text. The unstandardized metric is somewhat simpler when using a matrix algebra formulation.

Readers should be slowly weaned away from exclusive reliance on standardization. Other texts on this topic, both advanced and beginning, almost exclusively employ analysis using the raw metric. One may need to first translate some of these analyses to the standardized metric, but eventually one should be able to handle the problem entirely in terms of the unstandardized metric.

Remember to be flexible. The important thing is to learn from data. One must learn only with what one knows, but one must be willing to learn new methods in order to learn even more.

CONCEPTUAL ISSUES

Estimation versus Testing

Traditionally the major emphasis within the causal modeling literature has been on the *estimation* of structural parameters. An alternative and sometimes competing emphasis has been the *testing and comparison* of causal models. To some extent this distinction parallels the classic distinction between descriptive and inferential statistics.

The best way to illustrate the differences between the two emphases is to show how they differently handle overidentification. Let us say we have a model that is overidentified and there are two different researchers who approach it. The first researcher is interested in estimation. He or she wants the most efficient estimates of parameters. The researcher is overjoyed that the model is overidentified for the following reason. Since a given parameter estimate is overidentified, there are two or more estimates of the same parameter. The researcher can then pool these multiple estimates to obtain a more efficient estimate. There may be some technical problems on how to pool the multiple estimates (Goldberger, 1973), but it always involves a weighted sum of the estimates.

The second researcher is also overjoyed the model is overidentified, but for a very different reason. Since there is overidentification, it should be possible to find an overidentifying restriction. With a little imagination a statistical test can be found and it can be determined whether the overidentifying restriction is satisfied by the data. If the overidentifying restriction is not satisfied by the data, then the model contains a specification error.

The aims of the two researchers are not incompatible since it is always possible to test if the overidentifying restriction is met in the sample, and if it is, to pool estimates to obtain more efficient estimates of parameters. Unfortunately, though, some researchers fail to test overidentifying restrictions and others do not obtain efficient estimates of parameters.

The historical reason for the greater emphasis on estimation is due to the fact that the growth of structural modeling occurred first among econometricians. The typical data set they use is a time series with usually no more than 25 observations. It would seem that this would be an incredibly hostile environment for causal modeling to grow, but grow it did. Faced with the problem of serial dependency, the issue of estimation became paramount. Moreover, given the small sample sizes and very high multicollinearity, the efficiency of estimation is even more critical. Readers of econometrics texts often fail to realize the collinearity of economic data since usually only covariance matrices are given. For instance, for the inventory–demand model discussed by Goldberger (1964, p. 286) the median correlation is .92, and lowest correlation is .90. It is hardly surprising that econometricians worry about the efficiency of their estimates.

A second reason for the emphasis on estimation over testing is that econometricians were the pioneers of the theory of identification. The central question is whether a model is identified in the first place not whether it is overidentified. Thus, the focus was on identification and not overidentification.

A third reason for the emphasis on estimation within econometrics is the very natural metric that they possess: dollars. Economists, unlike psychologists, are unabashedly applied and as a matter of course want to know the effect in dollars and cents of the exogenous variables. Estimation of structural parameters is then very useful for economic policy.

Estimation is an important goal in causal modeling, but it should not blind one from considering the testing of models as also important. One should not neglect the testing of overidentifying restrictions. Also one should consider models in which none or few of the parameters are estimable, but the model is still overidentified. Examples of this are in Figure 3.6 and throughout Chapters 11 and 12. Normally, parameter estimates are complicated combinations of covariances but it may be that a simple comparison of two covariances or correlations reveals something very interesting about a model.

Still overidentifying restrictions are not the be all and end all of causal models. A major problem with an overidentifying restriction is that, if it fails, it may not be able to pinpoint where the specification error exists. Ordinarily if the restriction is not met, a whole multitude of specifications could be incorrect, either only one or all. Thus not meeting the restriction may not be sufficiently diagnostic about where the specification error is.

Also, even if the restriction is met the model is still not proved to be true. In the first place, the test of the overidentifying restriction may be of low power. For instance, if correlations are low, then the test of a vanishing tetrad is of low power. Second and even more fundamental, if the restriction holds, there are an infinite set of similarly overidentified models that would satisfy exactly the same set of restrictions.

We should never lose sight that causal models can never be proven to be true. Certainty must be left to the mystic. Causal models can be disconfirmed by the data but never is a *single* model confirmed, but rather a host of such models. No doubt that after finishing this text the reader has a sense of disillusionment. Before starting the text, it was perhaps hoped that one would learn magical methods to use correlations to test theory. Perhaps some hoped for a black box into which one would put the correlations and out would come the theory that explained the correlations. Such is not the case. It should now be clear that causal modeling requires the researcher to already know a great deal about the processes that generated the data. Knowledge does not grow in a vacuum.

There is still an advantage in having a model that is overidentified. A model should say not only what should happen with the data but also what should not. If a zero path or an overidentification follows from a model, then the path must be zero or overidentification must hold. If they do not hold then the model is misspecified. One can, of course, rescue the model by adding an additional parameter, but, hopefully, honest researchers will lessen their confidence in the model.

Campbell and Fiske (1959) make a valuable contribution in their discussion of the validity of constructs by introducing *convergent* and *discriminant validity*. Metaphorically one can test the convergent and discriminant validity of a structural model. Convergent validity refers to the fact that a model replicates across different samples, operationalizations of constructs, and estimation methods. Duncan (1975, pp. 152–158) provides a very useful discussion of the replication of estimates of a structural parameter across different samples. Just as important is discriminant validation. The model should imply very different things than alternative competing models. This will usually take the form of zero paths and overidentifying restrictions. Ideally data should be able to discriminate between alternative structural models. One can never rule out all possible models but we should attempt to rule out the strongest competing alternative models.

For correlational studies the classic Campbell and Stanley (1963) distinction between internal and external validity breaks down. To some extent internal validity can be defined as all the threats to causal inference that are controlled for by a true experiment. Thus by definition a correlational study lacks internal validity. It would then seem that the more appropriate goals of a correlational study are convergent and discriminant validity.

Model Specification

Two errors of model specification are common to beginning causal modelers. The first is the tendency to draw path models in which everything causes everything else. All variables are assumed to be measured with error and all the variables are involved in feedback relationships. When one reads theory papers on various topics, one usually finds a pseudopath diagram in which every variable causes the other. As the reader must now realize, although such a model may be theoretically interesting, it is most certainly underidentified; that is, it is empirically useless. The researcher and the theorists must learn to pare a theory down to a smaller number of paths. Such a simplification will involve distortion, but if the omitted paths are small then such distortion should be minor. Each of us should realize that all models are most certainly misspecified, but hopefully the specification errors are minor.

One common type of "overspecification" is measurement error in measured exogenous variables. Readers of Chapter 5 should not despair of identifying causal models when there are "errors in variables." If reliability is high the effect of measurement error will be small.

One still should be very careful to avoid an "ad hoc specification." An ad hoc specification is one that is chosen solely with the purpose of identification in mind. It strikes me that often the zero path assumption made in instrumental variable estimation is often made for the sake of expediency rather than being motivated by theory. One must avoid the trap of ad hoc specification.

Related to ad hoc specification is the need to respecify a model after estimation and testing. This is a normal practice and a researcher should not be embarrassed. What one should avoid in writing up and presenting results is discussion of only the final model as if it were the only model. The reader deserves to know and to be instructed about the initial model and the steps that were taken to obtain the final model. An excellent example of the development of a final model is contained in Duncan, Haller, and Portes (1971).

The second problem that beginners have with model specification is underspecification. The model is nicely overidentified but it is clearly unrealistic. Be realistic about all possible linkages in a model. Read the literature and know which paths have been found in the past. It is indeed useful to always start with (or work to) the simplest and most parsimonious model, but realize that the world is not as simple as we might like it to be.

In Search of High R^2

The traditional criterion for evaluating a regression equation is R^2. It is generally thought that the larger the R^2, the better the model is. It would be nice to be able to predict 100% of the variance, but this may not be feasible or even possible. As suggested in Chapter 1, one may take the ontological position that empirical phenomena are not perfectly predictable in both theory and practice. As a practical rule there often appears to be an upper limit to R^2.

In the first place the researcher should avoid questionable statistical methods that capitalize on chance. The R^2 should be adjusted downward to take into account any such capitalization. Moreover, there are three guaranteed methods of increasing R^2 that involve no real theoretical insight: using an alternative measure, a lagged value, and aggregation of units. The alternative measure method is simply to use another measure of the endogenous variable as an exogenous variable. In such a case the R^2 will be more a measure of reliability than predictability. A second and related method is to use a lagged value as an exogenous variable; if one is attempting to predict intelligence at age 10, use intelligence at age 9 as a predictor. The third method of achieving a high R^2 is to aggregate the data. Combine subjects into groups and make groups the unit of analysis. As a rule aggregated data yield a higher R^2 than individual level data. But beware of the ecological fallacy. The inferences for the aggregated data refer to the aggregated units and not the individual. The goal of causal modeling is not to maximize R^2 but to test theory. One would suppose that better theories have higher R^2 but this does not necessarily follow.

One can legitimately increase R^2 by having models that include unmeasured variables. For instance, let X_1 and X_2 be equally reliable indicators of a latent construct F with reliability a^2 and assume that F causes X_3. If $\rho_{13} = \rho_{23} = b$ then the multiple correlation squared of X_1 and X_2 with X_3 is

$$\frac{2a^2b^2}{1 + a^2} \qquad [13.3]$$

whereas the correlation squared of F with X_3 is b^2/a^2. Note for any value of b the correlation of F with X_3 is always greater than Equation 13.3 ($a^2 < 1$) and that for a fixed b, as a decreases the correlation of F with X_3 increases. Thus for a fixed set of correlations between indicators and an endogenous variable, the correlation of the factor with that variable increases as the reliabilities of the measures decrease!

Although a high R^2 is not the absolute goal, some correlation is necessary for causal analysis. Please take my advice and do not vainly attempt to salvage a matrix of essentially zero correlations. There are tests of significance that triangular and rectangular matrices have only zero correlations. Such tests should be used more often to discard useless data sets.

Neglected Topics and Procedures

This text is meant to be only an introduction to causal modeling. Although only introductory, it has taken a rather broad sweep. Some of these topics have been covered rather superficially and tangentially. The hope is that the broad exposure will whet the appetite for more. But remember, a little sip is dangerous, so drink deep.

There has been no extensive discussion of time series. This would be a text in and of itself. Also nowhere is the estimation method of generalized least squares. This estimation technique is useful for overidentified models, "seemingly unrelated equations," and time series. Not enough attention has been paid to various statistical assumptions (e.g., distributional) and their implications to statistical estimates. Finally and perhaps most seriously, the text has not employed matrix algebra. Thus, many of the methods discussed have not been adequately explained for the general case. However, this book is an introduction. The concern here is for initial comprehension of a simple case. Then once that is learned, the reader will be motivated to learn the general case. The book, to use a metaphor of Wittgenstein, is a ladder with which one can climb to new heights, but once one has climbed as high as the ladder will take one, the ladder must be cast aside.

Bibliography

Althauser, R. P. Multicollinearity and nonadditive regression models. In H. M. Blalock (Ed.), *Causal models in the social sciences*. Chicago: Aldine-Atherton, 1971.

Alwin, D. F. Approaches to the interpretation of relationships in the multitrait–multimethod matrix. In H. L. Costner (Ed.), *Sociological methodology 1973–1974*. San Francisco: Jossey-Bass, 1974.

Alwin, D. F., & Hauser, R. M. The decomposition of effects in path analysis. *American Sociological Review*, 1975, *40*, 37–47.

Alwin, D. F., & Tessler, R. C. Causal models, unobserved variables, and experimental data. *American Journal of Sociology*, 1974, *80*, 58–86.

Anscombe, F. J. Statistical analysis, outliers. In D. L. Sills (Ed.), *International encyclopedia of the social sciences*. New York: Macmillan & The Free Press, 1968.

Appelbaum, M. I., & Cramer, E. M. Some problems in the nonorthogonal analysis of variance. *Psychological Bulletin*, 1974, *81*, 335–343.

Bachman, J. G., & Jennings, M. K. The impact of Vietnam on trust in government. *Journal of Social Issues*, 1975, *31*, 141–155.

Ball, S., & Bogatz, G. A. *The first year of Sesame Street: An evaluation*. Princeton, N.J.: Educational Testing Service, 1970.

Berman, J. S., & Kenny, D. A. Correlational bias in observer ratings. *Journal of Personality and Social Psychology*, 1976, *34*, 263–273.

Birnbaum, M. H. The devil rides again: Correlation as an index of fit. *Psychological Bulletin*, 1973, *79*, 239–242.

Blalock, H. M. *Causal inferences in nonexperimental research*. Chapel Hill: University of North Carolina, 1964.

Blalock, H. M. *Theory construction: from verbal to mathematical formulations*. Englewood Cliffs N.J.: Prentice-Hall, 1969.

Bock, R. D. *Multivariate statistical methods in behavioral research*. New York: McGraw-Hill, 1975.

Bohrnstedt, G. W. Observations on the measurement of change. In E. F. Borgatta (Ed.), *Sociological methodology 1969*. San Francisco: Jossey-Bass, 1969.

Bohrnstedt, G. W., & Marwell, G. The reliability of products of two random variables. In K. F. Schuessler (Ed.), *Sociological methodology 1978*. San Francisco: Jossey-Bass, 1977.

Bonacich, P., & Kirby, D. Using assumptions of linearity to establish a metric. In D. R. Heise (Ed.), *Sociological methodology 1976*. San Francisco: Jossey-Bass, 1975.

Brewer, M. B., Campbell, D. T., & Crano, W. D. Testing a single factor model as an alternative to the misuse of partial correlations in hypothesis-testing research. *Sociometry*, 1970, *33*, 1–11.

Bryk, A. S., & Weisberg, H. I. Use of the nonequivalent control group design when subjects are growing. *Psychological Bulletin*, 1977, *84*, 950–962.

Calsyn, R. J. *The causal relation between self-esteem, locus of control, and achievement: A cross-lagged panel analysis*. Unpublished doctoral dissertation. Northwestern University, 1973.

Calsyn, R. J., & Kenny, D. A. Self-concept of ability and perceived evaluation of others: Cause or effect of academic achievement? *Journal of Educational Psychology*, 1977, *69*, 136–145.

Campbell, D. T. From description to experimentation: Interpreting trends as quasi-experiments. In C. W. Harris (Ed.), *Problems in measuring change*. Madison: University of Wisconsin, 1963.

Campbell, D. T. *The effect of college on students: Proposing a quasi-experimental approach*. Research Report, Northwestern University, 1967.

Campbell, D. T. Reforms as experiments. *American Psychologist*, 1969, *24*, 409–429.

Campbell, D. T. Temporal changes in treatment–effect correlations: A quasi-experimental model for institutional records and longitudinal studies. In G. V. Glass (Ed.), *The promise and perils of educational informations systems*. Princeton, N.J.: Educational Testing Service, 1971.

Campbell, D. T. Evolutionary epistemology. In P. A. Schlipp (Ed.), *The philosophy of Karl R. Popper*. LaSalle, Ill.: Open Court Publishing, 1974.

Campbell, D. T., & Erlebacher, A. How regression in quasi-experimental evaluation can mistakenly make compensatory education harmful. In J. Hellmuth (Ed.), *The disadvantaged child*, Vol. III. *Compensatory education: A national debate*. New York: Brunner-Hazel, 1970.

Campbell, D. T., & Fiske, D. W. Convergent and discriminant validation by the multitrait–multimethod matrix. *Psychological Bulletin*, 1959, *56*, 81–105.

Campbell, D. T., & Stanley, J. C. Experimental and quasi-experimental designs for research on teaching. In N. L. Gage (Ed.), *Handbook of research on teaching*. Chicago: Rand McNally, 1963.

Clyde, D. J., Cramer, E. M., & Sherin, R. J. *Multivariate statistical programs*. Coral Gables, Fl., University of Miami, 1966.

Cohen, J. *Statistical power analysis for the behavioral sciences*. New York: Academic, 1969.

Cohen, J. Partialed products *are* interactions; partialed powers *are* curve components. *Psychological Bulletin*, 1978 *85*, 858–866.

Cohen, J., & Cohen, P. *Applied multiple regression/correlation analysis for the behavioral sciences*. Hillsdale, N.J.: Lawrence Erlbaum, 1975.

Cook, T. D., Appleton, H., Conner, R. F., Shaffer, A., Tamkin, G., & Weber, S. J. *"Sesame Street" revisited: A case study in evaluation research.* New York: Russell Sage, 1975.

Cook, T. D., & Campbell, D. T. The design and conduct of quasi-experiments and true experiments in field settings. In M. D. Dunnette (Ed.), *Handbook of industrial and organizational research.* Chicago: Rand McNally, 1976.

Costner, H. L. Theory, deduction, and rules of correspondence. *American Journal of Sociology,* 1969, *75,* 245–263.

Costner, H. L., & Schoenberg, R. Diagnosing indicator ills in multiple indicator models. In A. S. Goldberger & O. D. Duncan (Eds.), *Structural equation models in the social sciences.* New York: Seminar, 1973.

Crano, W. D., Kenny, D. A., & Campbell, D. T. Does intelligence cause achievement? A cross-lagged panel analysis. *Journal of Educational Psychology,* 1972, *63,* 258–275.

Cronbach, L. J., & Furby, L. How we should measure "change"—or should we? *Psychological Bulletin,* 1970, *74,* 68–80.

Cronbach, L. J., Gleser, G. C., Nanda, H., & Rajaratnam, N. *The dependability of behavioral measurements.* New York: Wiley, 1972.

Cronbach, L. J., Rogosa, D. R., Floden, R. E., & Price, G. G. *Analysis of covariance in nonrandomized experiments: Parameters affecting bias.* Evaluation Consortium, Stanford University, 1978.

Director, S. M. *Underadjustment bias in the quasi-experimental evaluation of manpower training.* Unpublished doctoral dissertation, Northwestern University, 1974.

Draper, N. R., & Smith, H. *Applied regression analysis.* New York: Wiley, 1966.

Duncan, O. D. Path analysis: Sociological examples. *American Journal of Sociology,* 1966, *72,* 1–16.

Duncan, O. D. Some linear models for two-wave, two-variable panel analysis. *Psychological Bulletin,* 1969, *72,* 177–182.

Duncan, O. D. Partials, partitions, and paths. In E. F. Borgatta & G. W. Bohrnstedt (Eds.), *Sociological methodology 1970.* San Francisco: Jossey-Bass, 1970.

Duncan, O. D. Unmeasured variables in linear models for panel analysis: Some didactic examples. In H. L. Costner (Ed.), *Sociological methodology 1972.* San Francisco: Jossey-Bass, 1972.

Duncan, O. D. *Introduction to structural equation models.* New York: Academic, 1975.

Duncan, O. D. Some linear models for two-wave, two-variable panel analysis with one-way causation and measurement error. In H. M. Blalock et al. (Eds.), *Mathematics and sociology.* New York: Academic, 1976.

Duncan, O. D., Haller, A. O., & Portes, A. Peer influences on aspirations: A reinterpretation. In H. M. Blalock (Ed.), *Causal models in the social sciences.* Chicago: Aldine-Atherton, 1971.

Fienberg, S. E. *The analysis of cross-classified data.* Cambridge, Ma.: MIT Press, 1977.

Fine, G. A. Psychics, clairvoyance and the real world: A social psychological analysis. *Zetetic,* 1976, *1,* 25–33.

Finney, J. M. Indirect effects in path analysis. *Sociological Methods & Research*, 1972, *1*, 175–186.

Goldberger, A. S. *Econometric theory*. New York: Wiley, 1964.

Goldberger, A. S. On Boudon's method of linear causal analysis. *American Sociological Review*, 1970, *35*, 97–101.

Goldberger, A. S. Econometrics and psychometrics. A survey of communalities. *Psychometrika*, 1971, *36*, 83–107.

Goldberger, A. S. Efficient estimation in overidentified models: An interpretive analysis. In A. S. Goldberger & O. D. Duncan (Eds.), *Structural equation models in the social sciences*. New York: Seminar, 1973.

Goodman, L. A. Causal analysis of data from panel studies and other kinds of surveys. *American Journal of Sociology*, 1973, *78*, 1135–1191.

Goodman, L. A. The analysis of systems of qualitative variables when some of the variables are unobservable. Part I—A modified latent structure approach. *American Journal of Sociology*, 1974, *79*, 1179–1259.

Gordon, R. A. Issues in multiple regression. *American Journal of Sociology*, 1968, *73*, 592–616.

Gorsuch, R. L. *Factor analysis*. Philadelphia: W. B. Saunders, 1974.

Guilford, J. P. *Psychometric methods*. New York: McGraw-Hill, 1954.

Hamilton, V. L. *The crime of obedience: Jury simulation of a military trial*. Unpublished doctoral dissertation, Harvard University, 1975.

Hargens, L. L., Reskin, B. F., & Allison, P. D. Problems in estimating measurement error from panel data: An example involving the measurement of scientific productivity. *Sociological Methods & Research*, 1976, *4*, 439–458.

Harmon, H. H. *Modern factor analysis*, 2nd ed. Chicago: University of Chicago, 1967.

Harris, R. J. *A primer of multivariate statistics*. New York: Academic, 1975.

Hauser, R. M. Disaggregating a social–psychological model of educational attainment. In A. S. Goldberger & O. D. Duncan (Eds.), *Structural equation models in the social sciences*. New York: Seminar, 1973.

Hauser, R. M., & Goldberger, A. S. The treatment of unobservable variables in path analysis. In H. L. Costner (Ed.), *Sociological methodology 1971*. San Francisco: Jossey-Bass, 1971.

Hays, W. L. *Statistics*. New York: Holt, Rinehart, & Winston, 1963.

Heise, D. R. Separating reliability and stability in test–retest correlation. *American Sociological Review*, 1969, *34*, 93–101.

Heise, D. R. Causal inference from panel data. In E. F. Borgatta & G. W. Bohrnstedt (Eds.), *Sociological methodology 1970*. San Francisco: Jossey-Bass, 1970.

Heise, D. R. *Causal analysis*. New York: Wiley-Interscience, 1975.

Huck, S. W., & McLean, R. A. Using a repeated measures ANOVA to analyze the data from a pretest–posttest design: A potentially confusing task. *Psychological Bulletin*, 1975, *82*, 511–518.

Humphreys, L. G. Investigations of a simplex. *Psychometrika*, 1960, *25*, 313–323.

Insko, C. A., Thompson, V. D., Stroebe, W., Shaud, K. F., Pinner, B. E., & Layton, B. D. Implied evaluation and the similarity-attraction effect. *Journal of Personality and Social Psychology*, 1973, *25*, 297–308.

Jaccard, J., Weber, J., & Lundmark, J. A multitrait–multimethod analysis of four attitude assessment procedures. *Journal of Experimental Social Psychology*, 1975, *11*, 149–154.

Jacobson, J. *The determinants of early peer interaction*. Unpublished doctoral dissertation, Harvard University, 1977.

Jencks, C., Smith, M., Acland, H., Bane, M. J., Cohen, D., Gintis, H., Heyns, B., & Michelson, S. *Inequality*. New York: Basic Books, 1972.

Jessor, R., & Jessor, S. *Problem behavior and psychosocial development*. New York: Academic, 1977.

Jöreskog, K. G. Estimation and testing simplex models. *British Journal of Mathematical and Statistical Psychology*, 1970, *23*, 121–145.

Jöreskog, K. G. Statistical analysis of sets of congeneric tests. *Psychometrika*, 1971, *36*, 109–133.

Jöreskog, K. G. A general method for estimating a linear structural equation system. In A. S. Goldberger & O. D. Duncan (Eds.), *Structural equation models in the social sciences*. New York: Seminar, 1973.

Jöreskog, K. G., Gruvaeus, G. T., & van Thillo, M. *ACOVS: A general computer program for the analysis of covariance structures*. Princeton, N.J.: Educational Testing Service Bulletin 70-15, 1970.

Jöreskog, K. G., & Sörbom, D. *LISREL III—Estimation of linear structural equation systems by maximum likelihood methods*. Chicago: National Educational Resources, 1976.

Jöreskog, K. G., & Sörbom, D. Statistical models and methods for the analysis of longitudinal data. In D. J. Aigner & A. S. Goldberger (Eds.), *Latent variables in socioeconomic models*. Amsterdam: North-Holland, 1977.

Kelley, T. L. Essential traits of mental life. *Harvard Studies in Education*, 1935, *26*, 146.

Kenny, D. A. Cross-lagged and synchronous common factors in panel data. In A. S. Goldberger & O. D. Duncan (Eds.), *Structural equation models in the social sciences*. New York: Seminar, 1973.

Kenny, D. A. A test of a vanishing tetrad: The second canonical correlation equals zero. *Social Science Research*, 1974, *3*, 83–87.

Kenny, D. A. A quasi-experimental approach to assessing treatment effects in the nonequivalent control group design. *Psychological Bulletin*, 1975, *82*, 345–362. (a)

Kenny, D. A. Cross-lagged panel correlation: A test for spuriousness. *Psychological Bulletin*, 1975, *82*, 887–903. (b)

Kenny, D. A., & Cohen, S. H. *Analysis of the nonequivalent control group design under sociological selection*. Unpublished paper, Harvard University, 1978.

Kerchoff, A. C. *Ambition and attainment*. Rose Monograph Series, 1974.

Kidder, L. H., Kidder, R. L., & Synderman, P. *Do police cause crime?A cross-lagged panel analysis*. Duplicated research report, Temple University, 1974.

Kruskal, J. B. Statistical analysis, the transformation of data. In D. L. Sills (Ed.), *International encyclopedia of the social sciences*. New York: Macmillan & The Free Press, 1968.

Lawley, D. N., & Maxwell, A. E. *Factor analysis as a statistical method*. London: Butterworths, 1963.

Lazarsfeld, P. F. Mutual relations over time of two attributes: A review and integration of various approaches. In M. Hammer, K. Salzinger & S. Sutton (Eds.), *Psychopathology*. New York: Wiley, 1972.

Levin, J. R., & Marascuilo, L. A. Type IV error and interactions. *Psychological Bulletin*, 1972, *78*, 368–374.

Linn, R. L., & Werts, C. E. Analysis implications of the choice of a structural model in the nonequivalent control group design. *Psychological Bulletin*, 1977, *84*, 229–234.

Lord, J. M. Large-scale covariance analysis when the controlled variable is fallible. *Journal of the American Statistical Association*, 1960, *55*, 307–321.

Lord, F. M. A paradox in the interpretation of group comparisons. *Psychological Bulletin*, 1967, *68*, 304–305.

Lord, F. M., & Novick, M. R. *Statistical theories of mental test scores*. Reading, Ma.: Addison-Wesley, 1968.

Magidson, J. Toward a causal model approach for adjusting for preexisting differences in nonequivalent control group situation: A general alternative to ANCOVA. *Evaluation Quarterly*, 1977, *1*, 399–420.

McNemar, Q. *Psychological Statistics*, 4th ed. New York: Wiley, 1969.

Milburn, M. A. *Process analysis of mass media campaigns*. Unpublished doctoral dissertation, Harvard University, 1978.

Nie, N. H., Hull, C. H., Jenkins, J. G., Steinbrenner, K., & Bent, D. H. *SPSS: Statistical package for the social sciences*, 2nd ed. New York: McGraw-Hill, 1975.

Overall, J. E., & Spiegel, D. K. Concerning least squares analysis of experimental data. *Psychological Bulletin*, 1969, *72*, 311–322.

Overall, J. E., Spiegel, D. K., & Cohen, J. Equivalence of orthogonal and nonorthogonal analysis of variance. *Psychological Bulletin*, 1975, *82*, 182–186.

Overall, J. E., & Woodward, J. A. Nonrandom assignment and the analysis of covariance. *Psychological Bulletin*, 1977, *84*, 588–594.

Pelz, D. C., & Andrews, F. M. Detecting causal priorities in panel study data. *American Sociological Review*, 1964, *29*, 836–848.

Peters, C. C., & VanVoorhis, W. R. *Statistical procedures and the mathematical bases*. New York: McGraw-Hill, 1940.

Porter, A. C., & Chibucos, T. R. Selecting analysis strategies. In G. D. Borich (Ed.) *Evaluating educational programs and products*. Englewood Cliffs, N.J.: Educational Technology Publications, 1974.

Rickard, S. The assumptions of causal analyses for incomplete causal sets of two multilevel variables. *Multivariate Behavioral Research*, 1972, *7*, 317–359.

Rosengren, D. E., Windahl, S., Hakansson, P. A., & Johnsson-Smargdi, V; Adolescents' TV relations: three scales. *Communication Research*, 1976, *3*, 347–366.

Rosenthal, R., & Rosnow, R. L. *The volunteer subject.* New York: Wiley-Interscience, 1975.

Rozelle, R. M., & Campbell, D. T. More plausible rival hypotheses in the cross-lagged panel correlation technique. *Psychological Bulletin,* 1969, *71,* 74–80.

Rubin, D. B. Assignment to treatment group on the basis of a covariate. *Journal of Educational Statistics,* 1977, *2,* 1–26.

Sartre, J. P. *Search for a method.* New York: Vintage Books, 1968.

Schwartz, E., & Tessler, R. C. A test of a model for reducing measured attitude–behavior discrepancies. *Journal of Personality and Social Psychology,* 1972, *24,* 225–236.

Sibley, L. *An empirical study of the effects of context and linguistic training on judgments of grammaticality.* Unpublished paper, Harvard University, 1976.

Smith, M. L., & Glass, G. V. Meta-analysis of psychotherapy outcome studies. *American Psychologist,* 1977, *32,* 752–760.

Snedecor, G. W., & Cochran, W. G. *Statistical methods,* 6th ed. Ames, Iowa: Iowa State University, 1967.

Spearman, C., & Holzinger, K. The sampling error in the theory of two factors. *British Journal of Psychology,* 1924, *15,* 17–19.

Stanley, J. C. Analysis of unreplicated three-way classifications, with applications to rater bias and trait independence. *Psychometrika,* 1961, *26,* 205–219.

Staw, B. M. Attitudinal and behavioral consequences of changing a major organizational reward: A natural field experiment. *Journal of Personality and Social Psychology,* 1974, *6,* 742–751.

Steiner, I. D. Perceived freedom. In L. Berkowitz (Ed.), *Advances in experimental social psychology,* Vol. 5. New York: Academic, 1970.

Suppes, P. *A probabilistic theory of causality.* Amsterdam: North-Holland, 1970.

Tatsuoka, M. M. *Multivariate analysis.* New York: Wiley, 1971.

Thistlethwaite, D. L. & Campbell, D. T. Regression-discontinuity analysis: An alternative to the ex post facto experiment. *Journal of Educational Psychology,* 1960, *51,* 309–317.

Tukey, J. W. Data analysis: Sanctification or detective work. *American Psychologist,* 1969, *24,* 83–91.

Tukey, J. W. *Exploratory data analysis.* Reading, Ma.: Addison-Wesley, 1977.

Tyron, R. C., & Bailey, D. C. *Cluster analysis.* New York: McGraw-Hill, 1970.

Weiss, C. H. *Evaluation research.* Englewood Cliffs, N.J.: Prentice-Hall, 1972.

Werts, C. E., & Linn, R. L. A general linear model for studying growth. *Psychological Bulletin,* 1970, *73,* 17–22.

Werts, C. E., Linn, R. L., & Jöreskog, K. G. A congeneric model for platonic true scores. *Educational and Psychological Measurement,* 1973, *33,* 531–544.

Werts, C. E., Jöreskog, K. G., & Linn, R. L. Comment on the estimation of measurement error in panel data. *American Sociological Review,* 1971, *36,* 110–112.

Werts, C. E., Rock, D. A., Linn, R. L., & Jöreskog, K. G. Comparison of correlations, variances, covariances, and regression weights with or without measurement error. *Psychological Bulletin*, 1976, *83*, 1007–1013.

Wiley, D. E., & Wiley, J. A. The estimation of measurement error in panel data. *American Sociological Review*, 1970, *35*, 112–117.

Williams, E. J. The comparison of regression variables. *Journal of Royal Statistical Society Series B*, 1959, *21*, 396–399.

Wright, S. Correlation and causation. *Journal of Agricultural Research*, 1921, *20*, 557–585.

Yee, A. H., & Gage, N. L. Techniques for estimating the source and direction of causal inference in panel data. *Psychological Bulletin*, 1968, *70*, 115–126.

Index